Innocent Experiments

Studies in United States Culture

Grace Elizabeth Hale, *series editor*

Series Editorial Board
Sara Blair, University of Michigan
Janet Davis, University of Texas at Austin
Matthew Guterl, Brown University
Franny Nudelman, Carleton University
Leigh Raiford, University of California, Berkeley
Bryant Simon, Temple University

Studies in United States Culture publishes provocative books that explore U.S. culture in its many forms and spheres of influence. Bringing together big ideas, brisk prose, bold storytelling, and sophisticated analysis, books published in the series serve as an intellectual meeting ground where scholars from different disciplinary and methodological perspectives can build common lines of inquiry around matters such as race, ethnicity, gender, sexuality, power, and empire in an American context.

Innocent Experiments

Childhood and the Culture of Popular Science in the United States

Rebecca Onion

THE UNIVERSITY OF NORTH CAROLINA PRESS

Chapel Hill

This book was published with the assistance of the
Authors Fund of the University of North Carolina Press.

Manufactured in the United States of America
Designed by Jamison Cockerham
Set in Arno Pro and Wylie WF type
by Tseng Information Systems, Inc.
The University of North Carolina Press has been a member
of the Green Press Initiative since 2003.
Cover illustration: Cover art from the April 18, 1953, issue of *Collier's* magazine.

Library of Congress Cataloging-in-Publication Data
Names: Onion, Rebecca, author.
Title: Innocent experiments : childhood and the culture of popular science in
the United States / Rebecca Onion.
Other titles: Studies in United States culture.
Description: Chapel Hill : The University of North Carolina Press, [2016] | Series:
Studies in United States culture | Includes bibliographical references and index.
Identifiers: LCCN 2016009979| ISBN 9781469629469 (cloth : alk. paper) |
ISBN 9781469629476 (pbk : alk. paper) | ISBN 9781469629483 (ebook)
Subjects: LCSH: Science — Social aspects — United States. | Science projects —
Social aspects — United States. | Science — Study and teaching — United States.
Classification: LCC Q175.52.U6 O55 2016 | DDC 303.48/30973 — dc23
LC record available at http://lccn.loc.gov/2016009979

Contents

Illustrations

Acknowledgments

First things first. At the University of Texas at Austin, Julia Mickenberg and Janet Davis were always ready with advice, reading recommendations, and productively challenging feedback. I have tried to repay Janet and Julia for their scholarly generosity in jam and preserves, holiday fudge, and links to funny webpages, but I don't think the debt will ever be settled. Jeffrey Meikle, Bruce Hunt, and John Hartigan offered valuable perspective on this project from their corners of the scholarly world, as well as some grounding for my wild flights of fancy. A few years ago, Janet recommended that I speak with Mark Simpson-Vos, of the University of North Carolina Press, about publishing this project; that turns out to have been a great idea. Thanks to Mark and to UNC Press's Lucas Church and Stephanie Wenzel for their belief in *Innocent Experiments*.

A number of organizations and individuals have provided me with the gift of financial support. The University of Texas at Austin's Donald D. Harrington Fellowship and William S. Livingston Graduate Fellowship, as well as a generous postdoctoral fellowship at the Philadelphia Area Center for the History of Science (now the Consortium for History of Science, Technology, and Medicine), gave me precious years made free for work. (A special thanks to the Consortium's Babak Ashrafi, an excellent scholar and facilitator of scholarship.) Grants from the University of Texas at Austin's School of Liberal Arts and Department of American Studies, the Chemical Heritage Foundation, the Children's Literature Association, the Friends of the Princeton University Library, and the Robert A. Heinlein Online Archives all provided short-term funding. At Ohio University, the history department has kindly extended credentials that give me access to books and databases; thanks to Chair Katherine Jellison for facilitating this crucial affordance. Friends Mary and Doc Hopkins, Mira Manickam, Julia Turner and Ben Wasserstein, Audrey Federman and James Joughin, Elanor Starmer and Kumar Chandran, and Rebecca and Jake Maine have allowed me to shelter with them on my various archives jaunts.

Archivists and librarians found materials for me, showed me their treasures, and supported me with their enthusiasm and ideas. Among those I thank: at the Smithsonian, Pamela Hansen, Peggy Kidwell, and Marcel LaFollette; at the Cotsen Children's Library at Princeton, Andrea Immel and Aaron Pickett; at the Strong National Museum of Play, Lauren Soldano; at the Chemical Heritage Foundation, Rosie Cook; at the Brooklyn Children's Museum, Allison Galland; and at the American Museum of Natural History, Barbara Mathe and Gregory Raml.

At conferences, on Twitter, and over e-mail, the following scholars shared ideas, references, drafts, and helpful critiques: Jill Anderson, Alice Bell, Robin Bernstein, Sarah Bridger, Victoria Cain, Sarah Anne Carter, Natalia Cecire, Miriam Forman-Brunell, Ivan Gaskell, Adam Golub, Bert Hansen, David Hecht, Cindi Katz, Shane Landrum, Al Martinez, Erika Milam, Rebecca Miller, Teasel Muir-Harmony, Philip Nel, Alex Olson, Nathalie Op de Beeck, Leslie Paris, Karen Rader, Anna Redcay, Peter Shulman, William Turkel, Shirley Wajda, and Virginia Zimmerman. The editors and anonymous reviewers of the *Journal of the History of Childhood and Youth*, *Isis*, and *American Periodicals* provided feedback that significantly improved sections of this finished book. Audiences at meetings of the American Studies Association, the Children's Literature Association, the Exploring Childhood Studies Conference at Rutgers-Camden, and the Society for U.S. Intellectual History heard portions of various chapters and asked good questions. The University of Texas at Austin's History of Science Colloquium, the Penn Workshop Series in History and Sociology of Science, Medicine, and Technology, and the history department at Case Western Reserve University invited me to give presentations on this material and offered thoughtful feedback.

Several reading and writing groups have provided essential space to discuss, kvetch, and process. At the University of Texas at Austin, I exchanged drafts, mulled over ideas, or moaned and groaned about edits with Gavin Benke, Becky D'Orsogna, Andi Gustavson, Josh Holland, Katherine Feo Kelly, Stephanie Kolberg, and Anne Helen Petersen. More recently, I had a wonderful visit with the Cornell University Historians are Writers! group; I must thank Amy Kohout and Aaron Sachs, among others at Cornell, for their hospitality and helpful comments. Ohio University historians Kevin Matt-son, Kevin Uhalde, and Jaclyn Maxwell have provided welcome fellowship in Athens. The members of TimWeb, an online karass I am lucky enough to have found, would prefer to remain anonymous; if I could say their names, I'd sing them to the rooftops.

In the last three years of this book project's life, I've been writing his-tory for the Web, most frequently for *Slate* magazine. The editors and writers at *Slate*—especially my direct editor, John Swansburg—have sharpened my thinking and made my prose so much better. I benefit greatly from associat-ing with this group of smart, pragmatic people. Editors for *Aeon Magazine*, the *Boston Globe*'s "Ideas" section, the *Virginia Quarterly Review*, the *Appendix Journal*, and other publications also did their best to help me find my voice as I moved into a new phase of my writing life.

My personal obligations are many. I thank my extended family of benevo-lent Harris-Dolben-Swiggett aunts, uncles, and cousins. Particular gratitude goes to Don and Martha Dolben, who have supported me materially and emo-tionally through this endeavor, and to my grandmother Barbara Harris, whose affection for history I clearly inherited. I thank my Onion aunt and uncle, Charlene and Jay (now gone), for support through words and deeds. I thank my sweet in-laws, Gloria and Reese Muntean. (Jim Muntean, we miss you.)

And what of my longtime friend-family, scattered as we are across the globe? Thank you Sarah Bennett, Sarah Bridger, Sarah Coombs, Mira Edmonds, Cristie Ellis, Audrey Federman, Marisa Giorgi, Michael Kavanagh, Mira Manickam, Raj Nath, Andy Pratt, Elanor Starmer, Julia Turner, and Mark Zumwalt. I hope to best you in a freezing-cold round of boot hockey some-time very soon.

And, of course, my beloved immediate family: brother, Josh, and his part-ner, Greta Mahowald, who get me thinking about food, shelter, and plumb-ing when I've been staring at a screen too long, and who gave us my sweet, indefatigable nephew, Odin. My wise sister, Sarah Onion Alford, her stead-fast partner, Jody Alford, and my lively and astute nieces, Wilhelmina and Josephine: I kiss you from afar. And my parents, Anne and Perry, who fed me books from the start, hugged me often, and made sure we always had family dinner. I love you. I love you!

Finally, and most important, my Nick. After they made you, they broke the mold.

Innocent Experiments

Introduction

A Curious Century

In February 2012, President Barack Obama hosted a science fair in the White House, where he was photographed with a fourteen-year-old contributor to the fair, Joey Hudy of Phoenix, Arizona. One image of the event captured the commander in chief's wide-eyed expression as he gleefully operated Hudy's invention, the "Extreme Marshmallow Cannon." This image had an afterlife, circulating on Facebook and Twitter (@karinjr, with a link to a *Huffington Post* article on the event: "In other 'Obama is Adorable News,' may I remind you of this? . . . You may say 'Aw' now"; @TJHtwits: "What a lovely man Obama is!"). The Obama reelection campaign recognized the image's power and created an animated GIF of the event of the cannon firing, posting it on the campaign's official Tumblr page.

   In this image, Obama performed excitement, curiosity, and wonder in the face of a child's innovation, emotions that he stressed in the speech he gave to the assembled children, their parents, and the press. The subtext of the image: science and technology, especially when practiced by glasses-wearing young-sters, have the capacity to render even such powerful, worried men as Obama momentarily carefree. And by his performance, he was supporting those same glasses-wearing youngsters in an appealing quest for knowledge. Like Neil DeGrasse Tyson of the American Museum of Natural History and Bill Nye, "the Science Guy," two science popularizers Obama lauded in his presenta-

President Barack Obama operates Joey Hudy's marshmallow cannon at the White House Science Fair, February 7, 2012. Kevin Lamarque/Reuters/Corbis.

tion, Obama showed through this performance that he has "dedicated himself to making science cool for young people."

As he modeled these positive emotions for the assembled group, Obama's speech contained a message: children should hardly need to be told to like science; such a liking was genetically coded into the American personality. "We're a nation of tinkerers and dreamers and believers in a better tomorrow," Obama told the group. "You think about our Founding Fathers — they were all out there doing experiments — and folks like Benjamin Franklin and Thomas Jefferson, they were constantly curious about the world around them and trying to figure out how can we help shape that environment so that people's lives are better. It's in our DNA."[1] In identifying curiosity and a public-minded dedication to manipulation of the physical environment as "part of our DNA," Obama collapsed biological and cultural heritage, claiming science-mindedness as part of an American intellectual tradition that manifests itself in the young of each generation, whether or not that generation is biologically descended from the actual Founding Fathers. By invoking a love of experiment as heritage, Obama also gestured at other tropes of American exceptionalism: individual excellence, balanced with concern for the welfare of a larger community; pragmatic approaches to problem solving; and a commitment to mobility and personal freedom.

President Obama is a twenty-first-century descendant of the string of twentieth-century American adults — politicians, authors, teachers, journalists, museum workers, artists, toymakers — who have celebrated American children's "natural" potential in the scientific fields. Decades of White House science fair photo opportunities, beginning in the 1940s with the advent of the Westinghouse Science Talent Search, show how the science-talented youngster has been a national icon — a symbolic catalyst for pride, much like war heroes, pilots who stick dangerous landings, or the football team that has won the Superbowl.

This pride coexists with half a century's worth of worry over the paltry quantity of children who end up excelling in science. Critics often critique kids' emotional attitudes toward science, technology, engineering, and mathematics (STEM), sighing that American children simply don't "love" science the way that they should (or, as Obama would put it, they do not see how science is "cool"). Science teaching in school during this century has been the object of continual renovation and reconstruction. Often, arguments over science education revolve around the same questions of emotional experience and its relationship to vocational commitment. Adults have worried that science in schools kills the joy of scientific practice, making things that are "naturally interesting" into boring chores.

Because of science promoters' encouragement of particular kinds of emotion, the parallel development of the science extracurriculum, which has often stepped in to offer utopian spaces of science practice supposedly driven by pure feelings of curiosity and wonder, offers an intriguing opportunity for cultural analysis. The history of the evolution of this extracurriculum shows how ideas about the nature of rational citizenship have mixed with changing concepts of the nature of education, the meaning of scientific practice, and the proper duties of childhood in creating children's scientific popular culture. Looking at this extracurriculum's style of emotional pedagogy can also expand our understanding of gender imbalances in scientific fields. While the experience of classroom learning in the twentieth century has often discouraged girls interested in scientific careers, the imagined maleness of the "fun science" offered alongside the formal curriculum may have done much to cement these exclusions.[2]

This history also shows how the belief in a child's joy in science, as practiced independently in leisure hours, has changed adult understandings of the meaning of science itself. Ideas and practices surrounding childhood have been central to the making of social meaning.[3] During the twentieth century in the United States, the folding of science into the set of activities deemed "typical" or "right" for children to practice in their leisure time meant that science would take on some of the universality and purity of childhood — or that science and childhood, as concepts associated with innocence, naturalness, purity, and timelessness, would mutually reinforce each other. This supposed relationship between science practice and the joys of childhood is so deeply ingrained in present-day American common wisdom that a quotation from Einstein, "The pursuit of truth and beauty is a sphere of activity in which we are permitted to remain children all our lives," appears on a page of a day-by-day Zen calendar; comparisons between children and scientists form the basis for blog posts, articles, and books;[4] and radio shows like *This American Life* and *Radiolab* explore kids' understandings of logic and numbers, with utterly adorable results.[5]

This trope of comparing children with scientists and scientists with children has generally been complimentary to both groups. But it is not a connection without ideology. The growth of this association between children, childhood, and science practice has been tied up in American attitudes toward some key aspects of modernity: innovation, futurity, "progress," and the social circulation of scientific knowledge. Sociologists of childhood have often noted that children occupy a unique place in the culture of modernity, signifying, as they do, both a repetition of the past and the promise of the future.[6] The close examination of the minds, habits, and tendencies of children—a project that was itself a product of modernity and was institutionalized through the fields of developmental psychology and educational theory in the early twentieth century—contains fears and hopes about appropriate social reproduction in the face of what was perceived as rapid social change. The child came to represent stasis and change at the same time, as traditions and new possibilities mingled. In promoting science play, twentieth-century American adults were often nostalgic about their own childhoods spent investigating and exploring. (As early as the postwar period, adults thought children were no longer as scientifically inclined as they had once been; our present-day lamentations have precedent.) In encouraging children to think "science is cool," adults coming to a personal understanding of the meanings of modernity could visualize a future regime of knowledge in which universal scientific literacy was natural, enjoyable, and fun.

In this book, I ask how larger changes—in industrialization, in understandings of human nature, in the professionalization of the sciences, and in America's position in the world—brought about an interesting shift in the ideologies surrounding children's science play. Before World War II, adults creating children's scientific popular culture relied on evolutionary models of history that saw white American children (in particular, boys) as the pinnacle of years of refinement in human thought. Starting in the Progressive Era, the question of which activities were "correct" for children was often answered by the theory of recapitulation, proposed primarily by German biologist Ernst Haeckel. This theory held that ontogeny (the development of an individual) would always recapitulate phylogeny (the development of a species). Stephen Jay Gould, in his comprehensive history of the concept's life in the scientific community, writes that this idea motivated the child study movement in the United States, for better or for worse, during the height of its popularity at the turn of the century. Academics invested in child study, most notably psychologist G. Stanley Hall, viewed recapitulation as their grand unifying theory, pointing to such "evidence" as children's love for water or for rhyming games to show that young people were moving through their "savage" or "tribal" phases, on their way to young adulthood, in which they would embody the medieval habit of mind: dreamy, mystical, and troubled by occasional bouts of madness.[7] Gould points out that although the idea of recapitulation was discredited in the scientific community by the turn of the century and the child study movement had produced internal critiques of its widespread use by 1910, the theory continued to have powerful influences within the theory of child development even up until the time of Dr. Spock, who wrote in 1968: "Each child as he develops is retracing the whole history of mankind, physically and spiritually, step by step."[8] Indeed, throughout the history of science education in the twentieth century, as we shall see, promoters of science-as-play have reiterated a belief in recapitulationism as supporting "evidence" for a range of arguments about the order in which scientific concepts are introduced and the type of children who can handle scientific knowledge. Adults were reassured that if given the correct tools, American children—especially American boys—would succeed in perpetuating the national project of accumulation of knowledge through avid scientific inquiry and practice. Because recapitulation theory relies implicitly on the existence of a hierarchy of races, its use in discussing children's knowledge created a space that effectively excluded nonwhite children from the most advanced ways of knowing.

During and after World War II, this confidence in white boys' ability to surpass their elders in scientific knowledge largely disappeared, with the previous array of racial reassurances vanishing and a new awareness of global competition (and the potential for devastating global nuclear conflict) rising in its place. Some adults in the postwar era blamed popular culture for this failure, folding fear of loss of scientific "manpower" into a larger moral panic over juvenile delinquency and social conformity by arguing that children's youthful enjoyment of science had been corrupted by a vapid and unsubstantial peer culture. Some, like science fiction writer Robert Heinlein, thought that boys who would otherwise be committed to STEM careers suffered at the hands of scientifically ignorant (and often female) teachers, authors of children's books, and librarians. Others, such as social critic Paul Goodman, posited that young people shied away from science because of the new association between science, technology, and the military. Still others believed that children failed to commit to science because of the new pressure to "achieve," which sapped the practice of its inherent fun; such proponents of this point of view as Frank Oppenheimer, founder of the San Francisco science museum the Exploratorium, favored a new approach intended to reconnect children and adults with what he saw as the universal joy of science practice. Part of the work of this book is to define this chronological change, asking how the shift in attitudes from prewar confidence to postwar anxiety might help us understand the complex present-day approach to the encouragement of science-as-play in American childhood.

POPULAR SCIENCE AND KIDDIE SCIENCE

My working definition of American "popular science"—a contested concept, to be sure—encompasses all the ways that scientific activity and knowledge operate in everyday life. Science lives on television, on the radio, in the movies, in newspapers and magazines and on the Internet, and in science fiction and popular nonfiction books. But I also include the kinds of everyday understandings of science that circulate in conversations between a doctor and an anxious patient, an environmental activist and an unbelieving family member, children watching wildlife television together or the members of a town planning board trying to decide whether to develop a site. Most important for this book, I think science education, both formal and informal, is a particularly potent site of popular science. Watson Davis, the head of Science Service, the science journalism nonprofit that founded the Science Talent Search, argued that children's participation in science fairs would make them into ambassadors, spreading the spirit of science to their elders. Davis had a strong incentive to advocate for this theory—Science Service was heavily invested in the usefulness of science fairs and searches. Whether or not adults have been effectively "educated" by their children's involvement with science—and that is probably impossible to measure—children's involvement with science has certainly constituted an emotionally charged site of contact between Americans and the project of "science," writ large. Whether through memories of their own youth or by witnessing the experiences of sons and daughters encountering science in school, Americans who did not grow up to follow careers in STEM fields will often think of science as the province of school-age children. Cultural tropes often tapped in media representations of childhood—the five-year-old's cute and irritating habits of asking about the color of the sky or obsessively memorizing names of dinosaurs; the elementary-school-age child's exploding-volcano science fair project—derive from and reinforce this association.

The knowledge that circulated in the basement chemistry lab, during the science fair, and through the children's encyclopedia assumed the form that it did because of the way that American adults perceived the duties, rights, and advantages of being an American child. In tying this study to a larger history of evolving attitudes toward childhood, I follow recent suggestions of historians of science interested in articulating the specific cultural parameters that define the circulation of scientific knowledge.[9] The nineteenth century, for example, saw the birth of organized scientific societies and professionalization in the United Kingdom and the United States—a development that called forth a new brand of "popular" science. In public spaces, particularly (though not exclusively) in the United Kingdom, the fruits of inquiry were offered for interested parties who were shut out from official scientific discourses.[10] The parlor, the gallery, the panorama, the exhibition, and the lecture were key sites of encounter with science for both children and adult laypeople, and the impresarios and promoters who made these encounters possible operated with a wide variety of motivations (from desire for profit to religious commitment to philanthropic zeal). In the twentieth century, as popular science in the United States expanded along with mass media and consumer markets, the child "public" came to mean something very specific to Americans; as Americans worried about, obsessed over, defined, and redefined the meaning of childhood, their creation of "science fun" for this particular "public" was shaped by their new understanding of the cultural significance of this phase of life.

The association between science and childhood in popular culture has serious implications for the public's perception of the nature of scientific activity. Contemporary observers of science in American culture, such as science writer Natalie Angier, bemoan the segregation of science activity in childhood and wonder why "childhood is the one time of life when all members of an age cohort are expected to appreciate science."[11] The common wisdom that children's investigations of the world around them are somehow akin to science practice—and, therefore, that children, if left to unfold "naturally," would of course translate the two-year-old's joy in pouring sand from one vessel to the other into the more directed and controlled "fun" of scientific investigation—contains within itself the converse idea that scientists are naturally childlike and unworldly, almost feral. Does the strong relationship between science and "childishness" imply that a love of science is impractical or ill-befitting an adult citizen? And what does the association between science and "innocence"—a characterization that contradicts the very real implications of scientific work for everyday life—do to forestall a real discussion of these implications? The close cultural ties created between childhood and science in the United States in the twentieth century—ties that imply an exclusive relationship between an age cohort and a complex set of intellectual practices—cast suggestive illuminations on the current landscape of public perceptions of science.

CHANGING CHILDHOODS

Shifting conceptions of the fundamental nature of childhood in American modernity have, of course, been a big part of the change in social function of scientific popular culture for children. Scholars in the history of childhood and in the loosely defined field of childhood studies have traced these changing conceptions, pointing to places where larger social factors have shifted the place of children within the culture. During the twentieth century, middle-class children of elementary school age definitively exited the world of wage labor while steadily gaining emotional power within their families and in the public sphere. Sociologist Viviana Zelizer has argued that the beginning of the twentieth century saw "an expulsion of children from the 'cash nexus,'" as most middle-class families began to see children as "an exclusively emotional and affective asset" rather than as workers contributing to the family income.[12] The process of what Zelizer calls "sacralization" of children's lives was both a product and a cause of large-scale changes in children's everyday experiences; these changes included the end of many forms of child labor, the segregation of children from adults in public spaces, the beginning of compulsory schooling, and the creation of markets for goods aimed exclusively at children. While middle-class children ceased to earn money, their families began to spend more on their clothes, toys, and books, and the practice of giving children an allowance to buy some of these goods themselves began to gain in popularity.[13]

The contours of children's popular science, as it took shape in the United States, were defined by these new social realities; as a result, popular science for American kids was decidedly middle class and white from the start. Children's scientific entertainments often cost money, and the children who were entertained by these sometimes-expensive toys and books were embodying a rational style of leisure, one that channeled their impulses toward novelty-seeking in favorable intellectual directions. Sometimes science play resulted in children earning money, through offering such services as testing water with a chemistry set or putting on a "chemical magic" show or through the winning of prizes or scholarships from chemistry set company contests or science fairs. These small "salaries" were entirely appropriate for the new model of childhood, which demanded that any "work" children performed be educative, enjoyable, and result only in financial compensations that would not wind up in the family coffers.

The vision of American children passing their newly abundant leisure time in science play appealed to American adults, as these adorable miniature investigations ideally combined productivity and pleasure, promising future success as well as providing immediate entertainment. This focus on child-hood scientific capacity was a particular early twentieth-century American permutation of what sociologist Chris Jenks has identified as an "Apollonian" view of the child—an understanding of childhood as a period of precivilized purity and, thus, extraordinary mental advantage. (Jenks juxtaposes this para-digm with a previous understanding of children as "Dionysian": uncontrol-lable, motivated by instinct, and in need of severe discipline, a paradigm that parents in some conservative sectors of American society still follow.)[14] In this new view, the child mind is pictured as unsullied, capable of connecting more perfectly to "truth" because it has yet to be clouded by the duties and obliga-tions of adulthood lived in modernity. Jenks describes the "Apollonian child" as "naturally good," with a "clarity of vision" that is seen as "the source of all that is best in human nature." Children participate in the essential activity of human life, and in a modern era, their potential, if properly educated, points the way to a new way of living. Jenks identifies this investment in shaping the child's intellect as one aspect of a new regime of social control. Following Foucault's analysis of the function of institutional power in both extraordi-nary and everyday situations, Jenks argues that whereas the Dionysian child is punished, the Apollonian child is watched: "The crudity of the old regime of control in social relations gives way to the modern disciplinary apparatus, the post-Rousseauian way of looking at and monitoring the child in mind and body." Throughout this book, I will show how adults providing opportuni-ties for children's science play strove to offer freedom for these small investi-gations, while also constantly watching children's progress and analyzing its meaning.[15]

Jenks posits that the "Apollonian child" is a unique product of moder-nity. In her work on "rationality" in education, Valerie Walkerdine, arguing in a similarly Foucauldian vein, writes that investment in the development of rational thinking in children is a fundamentally defensive impulse, a reaction to change that is intended to create a feeling of mastery for adults: "a fantasy of an omnipotent power over a calculable universe."[16] Gesturing toward educa-tional thinkers such as John Dewey and especially Jean Piaget, Walkerdine ar-gues that their philosophical investment in science and mathematics learning is a product of their interest in engineering a better social order: "The rational dream sought to produce children who would become adults without perverse pleasures. These are the hopes invested in the power of reason."[17]

If Jenks and Walkerdine view education in rationality as a "dream" of control, this book asks what happens when this idealized rationality is envisioned as intertwined with the freely determined realm of play. Anthropologist Sharon Stephens writes of childhood in modern culture: "The ideological construction of childhood as the privileged domain of spontaneity, play, freedom, and emotion could only refer to a society that contained and drew upon this private domain as the ground for public culture, discipline, work, constraint, and rationality."[18] Science play, as a form of directed inquiry (discipline, work, constraint, and rationality) that was also assumed to be "natural" to children's wants and needs, has represented an ideal synthesis between the rational and the spontaneous; by promoting these rational entertainments, adults could hope to encourage precious feelings of freedom, while also training children in regulated habits of mind that would serve them well in a modernized workforce.

The newly age-segregated form of public science culture for American children in the twentieth century—science presented in a "children's museum," in the toy shop, or in science fiction intended only for young people— was a function of an increasing tendency toward a defined and child-specific culture. Geographers of childhood have pointed out that the twentieth century, when most children in the United States left the workplace and entered the classroom, saw the development of an increasingly regulated and planned set of child-specific spaces.[19] While nineteenth-century popular science in the United Kingdom and the United States often addressed itself to a multigenerational audience, as lyceums, museums, and libraries sought to reach adults and children alike, the science toys, museums, and books I look at in this book were the product of new differentiations between culture "appropriate" for children and for adults.[20]

In the United States in the twentieth century, adult encouragement of children's science play included careful cultivation of the perceived "natural" curiosity of childhood. Twentieth-century children's scientific culture departed from the morality of the children's science books published in the United Kingdom during the eighteenth and nineteenth centuries; in these, pedagogy in scientific matters was also meant to impart religious values, as in the volumes called "scientific catechisms," which took the form of a religious text so as to teach scientific "truths."[21] Religious imperatives for scientific learning dropped by the wayside; while some nature study texts in the early part of the century still tried to associate learning from "nature's book" with getting closer to the divine, most of these twentieth-century cultural objects relied on a vague invocation of the "joy of learning," letting inquiry, faithfully pursued, substitute for the previously privileged virtues of obedience, modesty, and service.

Nature study's emphasis on the emotional experience of encounters with nature, as well as its incorporation of literature and art into pedagogy, were a bridge between the eighteenth- and nineteenth-century mode of promoting childhood scientific activity and the twentieth century's new values. Significantly, the vogue for nature study drew criticism from some practicing scientists. In 1899, psychologist Edward Thorndike decried this method of pedagogy as "sentimental" and argued that curiosity, not affection, would drive scientific investigation. "Not the girl who dearly loves her doll, but the one who cuts it open to see its insides, is likely to be an investigator of human physiology," Thorndike wrote. "The boy who collects moths, who steals birds' eggs, who pokes the unlucky crab over onto its back and in fascination watches his uncomfortable efforts to right himself, who takes his toy animals apart to put them together again, is nearer the scientific pathway than the noble product of sentimental nature study who loves the worms and cares for the dear plants."[22]

Thorndike's argument — made from the point of view of a scientist claiming personal experience and authority to speak — privileged a child's amoral curiosity over the types of nurturing investigations of nature promoted by others working in the fields of nature study and humane education.[23] Thorndike's salvo, published at the beginning of a century of science promotion, is a harbinger of the subtle and explicit gendered rhetoric that would define the conversation around children's scientific impulses. While Thorndike includes a girl in his comparative scenario, he locates her inside, with a doll, rather than outside, with worms and crabs; this already limits her scope of action. The scene of a (supposedly nurturing) girl disemboweling her doll would have been significantly more upsetting and strange to his readers than the parallel one of a boy provoking animals and stealing eggs; this "bad" boy was a familiar and beloved figure in literature at the end of the nineteenth century.[24] By naming these familiar "bad" behaviors as scientific, Thorndike claimed the impulses of science for the world of boyhood.

The books, toys, and museums Americans produced for children in the twentieth century encouraged the value of curiosity; they were also invoking a more desirable alternative to the consumption-oriented child, addicted to the buying and owning of goods. The twentieth century saw the entry of the child into the marketplace. Before this time, historian Gary Cross argues, young people were "repositories of received learning and tradition"; the inclusion of children in the market of consumer culture meant "adults accepted change by giving novelty to their offspring."[25] While some adults worried that children's new relationship to the market would leave them hopelessly strung out on novelty, framing children's curious relationship with the world as scientific, rather than insolent or greedy, allowed an acknowledgment of "modern children" as relentlessly interested in the new, while locating these qualities as positive and generative rather than transgressive.[26] This relocation represented a shift from earlier perceptions of the value of inquiry. In her cultural history of curiosity in the early modern period, literary scholar Barbara M. Benedict describes curiosity as profoundly troubling to the established order. Benedict calls curiosity a "cultural ambition," one that "resists control" and "threatens the status quo." For Benedict, modernity changes perceptions of the curious, as proof of science's power comes to light: "[Curiosity] comes to define the modern personality: the upstart."[27] According to philosopher and educationist John Dewey, whose ideas informed so much of the culture I examine in this book, the raw resource necessarily to train scientific thinking was curiosity—a trait which children had in abundance but which adults often lacked.

Dewey's discussions of curiosity show how profoundly twentieth-century expectations of childhood had incorporated this previously dangerous quality and transformed it into a positive good. Childhood habits of questioning, Dewey argued in his 1910 exposition of the scientific method, *How We Think*, are at the heart of the *feeling* of science: "In the feeling, however dim, that the facts which directly meet the senses are not the whole story, that there is more behind them and more to come from them, lies the germ of intellectual curiosity."[28] Dewey warned that once children became adults, they lose this "germ." Citing Sir Francis Bacon, who wrote in the *Novum Organum* (1620) that we "must become as little children in order to enter the kingdom of science," Dewey reminded his readers that there is an "open-minded and flexible wonder" inherent in childhood, an "endowment" easily lost in adulthood: "Some lose it in indifference or carelessness; others in a frivolous flippancy; many escape these evils only to become incased in a hard dogmatism which is equally fatal to the spirit of wonder."[29] Like Dewey, many of the promoters of science play who appear throughout this book have held that children were closer to "the kingdom of science" in spirit than were adults. The history of encouragement of children's playful scientific inquiry is a contradictory one, in which adults both seek to direct the "magic" of children's constant questions about the world into productive channels and encourage them to retain their originality and freshness of observation.

As is widely recognized by practitioners of the history of childhood,[30] finding evidence of children's reactions and contributions to culture can be difficult, bordering on impossible; adult archives do not often contain records of children's thoughts, and even when they do, because of the power differential between children and adults, it is hard to separate a "true" reaction from a statement given in order to please an adult inquirer. However, I have tried whenever possible to include such input from child audiences as I can. I have found children's voices in the archive of the Brooklyn Children's Museum, where child patrons published a series of periodicals documenting their activities at the museum; in the letters written to the *Chemcraft Science Club Magazine* from young people who had formed their own science clubs; and, at a remove, in oral histories documenting twentieth-century scientists' childhood relationships with chemistry sets, science-themed movies and television, and science fiction.

ALL-INCLUSIVE WONDER?

Although American adults often naturalized science play as a universal mode of engagement, fun for all modern children, the whiteness and maleness of the children depicted in encyclopedias, toys, and books implied that scientific hobbies were represented as the property of the privileged. Throughout my examination of the formation of this popular scientific culture, I found very few depictions of children of nonwhite ethnicities "doing" science; the few representations that I did find were incidental, appearing in photographs of young museum patrons, rather than intentional or aspirational, as would be illustrations in science books or on chemistry set box tops. This was particularly true in the pre–World War II era. I show how these exclusions were related to a perception that the ability to think scientifically was an evolutionary gift—an inheritance given to "American" (white) children. The obliteration of this understanding in the postwar era was part of the anxiety surrounding such efforts as the Science Talent Search; I discuss how a new rhetoric of meritocracy borrowed from and adapted the racism of earlier decades. This is particularly interesting in light of the shift made in other realms of children's culture toward inclusion; as Julia Mickenberg points out in her work on children's literature and radical politics in the postwar era, many authors of children's nonfiction after the war made a conscious effort to include illustrations depicting nonwhite children in science books and to explicitly address questions of racism.[31]

The presence of girls in this scientific popular culture is especially important to examine, as the official curriculum and the structure of the scientific profession often discouraged girls from committing to scientific careers. White girls who do appear in scientific popular culture often manifest as allies of science; they are spectators, easily wowed by their brothers' or male friends' proficiency in manipulating chemicals to give a magic show, or easily distracted counterpoints to the engaged boy hearing informal lessons from an uncle or other adult. Girls, this culture implied, were engaged in science insofar as it was spectacular, provided visual stimulation, or contributed to consumer culture. Historians of education and historians of science have shown how women and girls were shut out of scientific discourse at a number of educational and professional levels; this book illustrates how these foreclosures of interest also occurred in popular culture.[32]

Gender and power within children's popular scientific culture were also intergenerational matters. As women were increasingly shut out of the science classroom, they receded into the background, less commonly appearing as figurative science instructors in children's popular culture. The end of the powerful nature study movement in the 1930s marked a closing of opportunities for female science teachers in the schools; prior to that date, women interested in science, shut out of careers in research, would often turn to school-teaching in nature study. Prior to the twentieth century, fictional mothers and "maiden aunts" "were remarkable sources of information" to fictional children inquiring about the world around them.[33] By the postwar period, male science teachers were preferred, and this shift was reflected in the popular culture of the time. During the postwar period, moral panics over juvenile delinquency contained speculation that boys would go astray because their fathers were either absent or effectively castrated by excessive female agency within the domestic setting. The panic over a lack of scientific manpower took a similar shape, with onlookers speculating that boys lost their natural affection for science when women failed to teach them correctly.

The chapters of this book follow evolving ideologies and expressions of the ideal of science play across the cultural locations where they appear. I have selected sites of children's culture whose creators perceived themselves as innovative in their time. Chapter 1 begins in Brooklyn, in the first few decades of the twentieth century, when the progressive educators who founded the Brooklyn Children's Museum created a vision of the museum community as scientific utopia, populated by adorable middle-class inquirers. Chapter 2 looks at science on the market and in the home during the interwar years, when the sale of chemistry sets and reference books that promised connection to a larger industrial world inspired the building of boys' basement laboratories—sites that would later take on mythic proportions in the imaginations of critics panicking over the loss of "fun science" in children's lives. The postwar efforts of the Science Talent Search to define young scientists as both mainstream and exceptional, at a time when science promoters struggled to understand why children were not sticking with STEM, are the basis for Chapter 3. At the same time, I argue in Chapter 4, postwar juvenile science fiction's enraptured depictions of space travel and the emergence of the cultural figure of the "space cadet" aligned science play more firmly with the adventuresome qualities of boyishness. Chapter 5 looks at a 1970s reclamation project, the Exploratorium in San Francisco, a museum that tried to put children and adults in touch with their inner scientist by associating scientific discovery with the freedoms of childhood. I finish with some reflections on the meaning of this history, asking how we can dig deeper beneath the truism "Kids should love science!" and find useful cultural meanings.

FORWARD TO THE FUTURE

Understood in context, Obama's delighted reaction to young Joey Hudy's marshmallow cannon, coupled with his impassioned arguments for an increased commitment to STEM education, is the latest in years of adult expressions of mingled celebration and anxiety. The apparatus itself—a machine of war, designed to loft projectiles of a soft, innocent, sweet substance—embodies the paradoxes inherent in designing a science that is "fun." Adults have promoted science for children as a matter of simple enjoyment, stripped of its moral valence by the association with innocent childish play; this presentation, while intended to provoke children into "falling in love" with science, also denies both the difficulties of science practice and the complexities of science's operations within society. In looking at and creating images of children "doing" science, critically assessing the state of children's sense of wonder and discovery, and devising new forms of activities intended to fan the flames of scientific inquiry, those interested in promoting science as play also made arguments about gender, privilege, public space, and power.

Wonder House

The Brooklyn Children's Museum as Beautiful Dream

In a 1908 article in *Popular Science*, Anna Billings Gallup, the curator of the Brooklyn Children's Museum, described the nine-year-old institution as a paradise for scientifically minded city children: a research lab, library, and clubhouse, packaged in an enchanting old Victorian mansion. Gallup, who joined the museum's staff in 1903, was to spend thirty-five years as its head. With a bachelor of science degree in biology from MIT, she became an educator who taught biology for four years at the Hampton School.[1] Despite her experience in a more traditional classroom environment, she saw the museum, an offshoot of the Brooklyn Institute of Arts and Sciences, as a superior alternative to school. "In the absence of official relations with public or private schools the museum makes no demands on its visitors," she wrote. "It offers its privileges free to children of all ages and leaves each one to choose his own method of enjoyment."

Gallup followed this remarkable proclamation of childhood liberation with a paragraph describing the range of sedate activities visitors might pursue within the walls of the refurbished mansion in Brooklyn. "Whether he copies a label," she wrote, "reads an appropriate quotation, talks about the group of muskrats with his playfellows, spends an hour in the library or listens to the explanation of the museum 'teacher,' who gladly answers his questions and tells him stories, matters but little so long as the effect of his visit is to enhance his love for the best things in life."[2]

In the distance between the wide-open possibility of the museum ("no demands") and her actual expectations for the visitors' experiences, there is a subtle rhetorical gap between the ideal and the real. A child could do whatever he wanted, so long as "whatever he wanted" was to pursue a positive mode of inquiry, approved by educational theory: copying, reading, conversing, researching, listening. Gallup, like other progressive educators, believed that children's minds — if allowed to develop naturally — would gravitate toward these productive modes of inquiry. This museum, even more than a progressive school, would prove that the bent of children's minds was positive — and, necessarily, scientific. Yet this "natural development," paradoxically, required much attention and encouragement from adults.[3]

Why was Gallup writing about a children's museum in *Popular Science*? While the Brooklyn Children's Museum was technically a "children's museum" — a designation which does not inherently predict the teaching of science — children's education in various branches of science took pride of place in the museum's slate of activities, and the museum's founders and curators often gestured toward children's scientific practice and later scientific accomplishments in their public representations of their museum's work. The museum contained "departments" of botany, zoology, geology, meteorology, geography, and history; there was also a library stocking textbooks, popular science volumes, and magazines, including *National Geographic*, *Nature*, *Bird-Lore*, and some more ambitious fare, including the *American Journal of Science* and the *Journal of Applied Microscopy*.[4]

In invoking the mentally and morally improving joys of the natural world to readers of *Popular Science* circa 1908, Gallup was repeating familiar common wisdom. In the second half of the nineteenth century, amateur natural history and specimen collecting became a mainstream pastime for American adults, who touted the benefits of this kind of connection with nature: physical vigor, renewal of religious sentiment, healthful recreation. Numerous scientific societies concentrated individual interest in collecting through local civic associations, and books and magazines, produced and distributed through the newly powerful publishing industry and postal service, advised people in the best ways to collect flora and fauna and trumpeted the advantages of the natural history habit. Natural history museums, funded by Gilded Age wealth, run by professional naturalists, and attended by interested amateurs (who occasionally contributed to their collections) grew in number; the American Museum of Natural History was founded in 1869, the Field Museum of Chicago in 1893, and the Carnegie Museum in Pittsburgh in 1895.[5]

Children participated in the late-nineteenth-century natural history craze, reading books and magazines that advised them on how to collect and incorporated collecting into stories, as well as taking field trips with those unusual teachers who happened to take the initiative to lead them.[6] Theodore Roosevelt colorfully remembered his own young exploits in the pages of the American Museum of Natural History's magazine, *American Museum Journal*, in 1918, recalling the lessons in taxidermy his father arranged for him and self-deprecatingly describing the "worthless" observations he made of local birds in Egypt and Palestine as a fourteen-year-old. Roosevelt wrote that he was no different from any "ordinary boy who is interested in natural history"—a description that shows how common the hobby was at the time (or, at least, how common Roosevelt believed it to be).[7]

In the early twentieth century, the popularity of natural history as an adult pastime diminished. Anticruelty campaigns gave the nineteenth-century tradition of specimen collection an unsavory reputation; the use of specimens for home decoration went out of fashion as a less cluttered design aesthetic came to dominate popular taste.[8] An urbanizing population had fewer places in which to collect. And if the Gilded Age was a time of prodigious museum building, in the early twentieth century American museums pivoted from serving as places of active scientific practice toward advancing educational missions. Children's presence in museums—on school visits, in after-school programs—became a commonplace rather than a rarity, and some museums began to maintain collections that could be lent out to classrooms.[9] The Brooklyn Children's Museum was founded in 1899, after one of the curators at the Brooklyn Institute of Arts and Sciences visited Europe and was impressed by exhibits at the Manchester Museum in England, which seemed to appeal to many children. Children's museums housed in separate buildings were a particularly American phenomenon; they were also a particularly Progressive phenomenon, built on a belief that children needed a separate sphere where they could learn best. Brooklyn led the way in supporting a children's museum; Boston followed (1913), then Detroit (1917) and Indianapolis (1925).[10]

In the first decades of the twentieth century, museums with child patrons were responding to and working with the nature study movement, a complex and widespread curricular initiative that gained authority at the turn of the twentieth century and retained primacy through the 1930s. The movement focused on introducing students in elementary schools to principles of scien-

Isabel L. Whitney's Brooklyn Children's Museum seal.
Children's Museum News, December 1924.

tific observation through investigations of their local environments. Drawing from the tradition of nineteenth-century natural history, with its emphasis on "studying nature, not books" (in the words of Harvard's influential geologist and paleontologist Louis Agassiz), nature study amplified other ideas inherent in progressive education, including an emphasis on allowing children to form their own understandings through observation. Nature study drew from the anxieties of its time, manifesting a distinct critique of urban modernity in its advocacy of contact with nature.[11] The movement had allies in the major schools of education, supported a national society and a journal, and at its height created careers for many teachers as nature study specialists. Gallup was a founding member of the major professional group of nature study educators, the Nature-Study Society, and often wrote about the Children's Museum's activities in the society's journal.[12]

"ALERT FROM TOES TO CROWN"

The Brooklyn Children's Museum seal, created in 1924 by Isabel Whitney, was meant to represent Ariel, "the sprite or spirit of childhood . . . clothed with light, the irradiant power of the universe . . . encased in a star." Whitney wrote that she struggled to represent the concept of the museum, thinking that "such an advanced ideal as the Children's Museum must needs be represented by the pictorial language of modernism" but adding that "today is an outgrowth of yesterday," and so "recognized symbols of the past must help in its expression." She therefore settled on the shining child, arms and legs outstretched, a beacon totally open for inspection.

Both Whitney and Anne Lloyd, the author of a commemorative poem titled "In the Children's Museum" printed in the museum's journal, *Children's Museum News*, a few years earlier, participated in the Children's Museum's self-articulation as a safe, cozy, familial place, where children would enter a natural paradise of education. This place, full of children paying the right kind of attention to the right things, was also very pleasant for adults to regard.

> I heard a happy humming
> As though a swarm of bees
> Over a new-found garden
> Were voicing ecstasies.
>
> It came from eager children
> Who thronged upstairs and down,
> Discovering fresh wonders,
> Alert from toes to crown.
>
> They listened to a legend,
> And joined in nature games,
> Calling the bugs and beetles
> By learned Latin names.
>
> They buzzed about strange countries,
> They burrowed deep in books,
> And graced the maps and pictures
> With rapt and reverent looks.

America extended
Her arms to every child,
And little foreign faces
Looked up at her and smiled.

The air was warm with welcome,
They felt as free to roam
Through each enchanted chamber
As if they were at home.

And many a drop of nectar
Their young souls stored away
To make a golden honey
To sweeten life someday.[13]

In their representations of the learning child, Children's Museum educators such as Lloyd combined a sentimental perspective with the rapt commitments of progressive education. Although pragmatism, with its disavowal of authority figures, could be considered to be full of what Max Weber called "disenchantments," pragmatists found new enchantments in idealizing all things "modern."[14] John Dewey, pragmatist and educational philosopher (and mentor of at least one museum staff member, Louise Condit), believed that by teaching children to think scientifically, the teacher not only exploited natural tendencies of childhood but also empowered the modern individual. Progressive interest in childhood learning resulted in enchantments with childhood and science, investing the learning child with shining possibility: the ability to carry the human race further along the road to full understanding and mastery of the external environment.

Indeed, Dewey's modern child was one who could foresee future events, manage intuitive responses, and plan rational courses of action, all using scientific thinking. The difference between this (white American) child and the "savage" of recapitulation theory was that the child represented newness and possibility, not a rote repetition of history. John Dewey's daughter Evelyn Dewey wrote rapturously of "the modern child" in 1934: "He is set amidst resources greater than Solomon dreamed of. Science has given him the wings of the eagle, the fins of the fish, long-distance eyes and ears so that he masters space and time. Matter has become his plaything. Running it through the mold of his imagination he changes its form at will. He inherits the intimacy of the stars and even dreams of transcending the boundaries of the planet. He is the last throw out of the Pandora's box of civilization."[15] For this child of modernity, who looks very much like "Ariel" of the Children's Museum and whose (male) body occupies the very upper rung on the ladder of civilization, the mastery of space and time has provided new "playthings" for the "imagination." This child combines the inheritance of resources (a rich environment, naturalized in Dewey's prose as the province of all children), which include enhanced sensory awareness provided by technology, with his own imagination and dreams. The vision contradicts recapitulation theory. Dewey's child is not running through the same old phases of savagery on his way to adulthood but is instead a product of accelerated evolution; his instincts should be trusted, and his interests, indulged.

The Children's Museum educators described the museum's physical plant, often remarked upon in discussions of the museum project, in terms of this unique balance between past and future. During its first decades, the Children's Museum was housed in a Victorian mansion—the Adams House, in Bedford Park.[16] Gallup wrote that the house's "picturesqueness of situation" rendered it uniquely suited for this particular museum.[17] The Adams House became an integral part of the museum's self-image, appearing, for example, on the letterhead used by the Women's Auxiliary; the mansion was emblematic of the institution itself. Caroline Worth, writing for the journal *Childhood* in 1922, dramatized the founding of the museum as a natural regeneration of a structure that had once housed children as a family dwelling but that had become "grim and neglected, as would any home in which the sweetness of childhood failed to enter." The museum established itself almost of its own accord: "Hundreds of birds seemed to fly into the parlors of the old mansion, and to arrange themselves in glass cases," Worth wrote. "The dining room disappeared, likewise the kitchen and bedrooms. Butterflies of every hue joined with moths, beetles, and dragon-flies in forming an Insect Room. . . . The great family of children took possession of the building."[18] In a 1912 article about the museum in the *Independent*, journalist Sydney Reid nicknamed the place "The Children's Wonder-House," drawing attention to the imagined childish possession of the physical plant.[19]

The idea of a museum "just for kids" was particularly suited to this time and place. Twentieth-century adults increasingly created separate and idealized realms of childish existence; "islanded" places where children would be protected and encouraged also served as nostalgic locations adults could mythologize and idealize.[20] Children's Museum educators described the "island" of the museum as one that was possessed by children's intellects. Gallup wrote in a pamphlet rejecting a proposal to absorb the museum into the Brooklyn Institute of Arts and Sciences, "The child must feel that the whole plant is for him, that the best is offered to him because of faith in his power to use it."[21] The physical spaces reflected consideration for what an adult visitor from out of town called "short-legged humanity."[22] In 1919, the *Children's Museum News* noted that the children greatly appreciated the museum's acquisition of a set of small folding chairs and a low table: "No longer will children have to stand up or lean against the polished glass or lie on the floors on their stomachs when they laboriously print the names of the birds and butterflies they have drawn. It is such a comfort not to have to stretch up to the big tables made for big folks!"[23] When a class of librarians-in-training visited the museum in 1926, a librarian recounted a story that she said was typical of the children's proprietary attitude toward the museum: "One small boy, looking at the young women . . . scowled, and in no uncertain tone declared, 'This is a Children's Museum.'"[24]

The Children's Museum positioned itself as more appropriate for children than the other museums, zoos, and botanical gardens in the city and suggested that attendance could serve as preparation for later enjoyment of these more adult institutions. In 1919, for example, the *Children's Museum News* reported on a twelve-year-old boy from Coney Island who had read a book about nature activities in New York and had decided to make his way to all the institutions named.[25] Hitting upon the Children's Museum first, the *News* said, he realized that the museum could "serve as an introduction to the larger museums." "He understood the collections because they are simple, he found books to read about the collections; he could attend the lectures, make experiments, join the League, and work with the other boys for certificates of credit and a medal." In the end, the *News* opined, the boy would be able to return to his quest, more fully prepared to ingest adult material.[26]

During the Progressive Era, reformers dwelled on the plight of children growing up in urban centers, especially those who were working sons and daughters of immigrants.[27] Larger urban museums like the American Museum of Natural History served these immigrant populations, perceiving themselves as missionaries of nature, bringing kids in touch with flowers and rocks that they would not otherwise see, and imagining big changes would be wrought in young lives owing to the contact.[28] While the children who came to the Brooklyn Children's Museum were city children and admission to the museum was free (at least in 1912), they were not economically disadvantaged.[29] The neighborhoods bordering the museum—today known as Bedford-Stuyvesant and Crown Heights and primarily African American in ethnic makeup—were, during the early twentieth century, bedroom suburbs for middle- and upper-middle-class families.[30] Gallup wrote in 1908 that many patrons were introduced to the museum when they came "with their parents, or the family nurse." Gallup also referred to children "returning from country outings" in September and visiting the museum full of stories about how they had applied nature study to their experiences.[31]

The children who visited the Brooklyn Children's Museum were not "nature starved" like their big-city counterparts, nor did they arrive at its doors arrive empty-handed. The Brooklyn Institute conceived of the Children's Museum as a place where children could bring things they might find in their daily lives: "Boys and girls often find odd and curious animals, or plants, or minerals, about which they would love to know something." The scientific staff attached to the Children's Museum would be happy to identify this flotsam and jetsam for the museum's patrons, offering information about "its place, history, uses, name, and structure."[32] Some evidence suggests that children visiting the museum received supplementary materials from their parents; George Schoonhoven, a fourteen-year-old patron, reported to a local newspaper that his parents had bought him a microscope and a camera to help his research in natural history.[33]

A final clue to the middle-class status of museum boys and girls during the first quarter of the century lies in the stories the staff told about the success of their "graduates," many of whom returned from college to share their experiences with the curators (more on this below). Between 1910 and 1930, teenagers from the working class were far more likely to drop out of high school than their middle-class counterparts, and immigrant children (those who were born outside the United States or whose parents were) were the least likely of their cohort to end up attending any high school at all.[34] The museum's image as an "islanded" paradise of learning, where children's interests could shape scientific inquiry, relied on the middle-class status of its patrons. The "family" of children inhabiting the old Adams House was not pathetic, poor, or in need of discipline, and because of this, the Brooklyn Children's Museum could indulge in flights of fancy about an institution ruled by youth's desire to know.

ADORABLE STRIVERS

The children of the Brooklyn Children's Museum manifested an easy familiarity with science that adult onlookers found nothing less than cute. In her
1908 *Popular Science* piece, Gallup indulgently described the response of a
"two-year-old baby" to bird exhibits: "He can only say 'Chicken, chicken' as
he points his chubby fingers indiscriminately to the condor, albatross, and fla-
mingo."[35] That was a sweet deficit of understanding, one that growth—and
the museum—could fill. Describing the summer field activities of the Children's Museum League, a group of children who acted as fund-raising "boosters" for the museum, in November 1915, the *Children's Museum News* reported
that "before the summer was over, children of less than ten years of age were
talking quite freely about 'Pholus pandorus,' 'Hemaris thysbe,' and a hundred
other species of moths, butterflies, beetles, bugs, and grasshoppers."[36] The
News reprinted a letter from an eleven-year-old girl in 1926, asking "in scientific vein" whether there are people living on Mars. "I asked a college boy," said
Marie Sharpe, "and he said scientists say or imagine that there are long, thin,
straggly people there. But it seems impossible for people to live on or in a star.
. . . I should think if there were, it would be such a weight on the star that it
would fall to earth."[37] Marie's question indulged adult ideas about the fanciful
imaginations of children, combined with admiration at her attempt to sort her
fancy out "scientifically." The fact that Marie interrogated a "college boy," who
drew his answer from a confused mishmash of outdated science and science
fiction (and who may have been pulling her leg), made the joke even funnier.

The idea that younger children's participation in science was an occasion for gentle adult amusement was echoed and reinforced in the images
that accompanied *Children's Museum News* articles. During the early twentieth century, as cameras became easier to acquire and photographs easier
to reproduce, children—both advantaged and disadvantaged—took center
stage in a growing proliferation of photographic representations.[38] Often,
these images—alarming photos of blighted kids stranded in industrial jobs;
dreamy portraits of children of privilege in domestic settings; dutiful docu-
mentations of the work of schools and organizations—tried to depict children's educational development (or lack thereof) visually. Many expositions
and conventions during the years between the Civil War and World War II
featured photographs of children, often as part of exhibits designed to present the work of educational institutions and reform movements.[39] Expositions often featured actual children as well, so that the adult viewer could see

education occurring on film and in the flesh; at the 1904 St. Louis World's Fair, for example, a visitor could observe elementary school classrooms; at the Pan-Pacific Exposition in 1915, a Montessori kindergarten was held in an amphitheater designed to accommodate adult visitors.[40]

Photographs printed in the *Children's Museum News* responded to the expectation that adults would like to watch cute children learn. This era saw a proliferation of pictorialist images of childhood made by American photographers such as Gertrude Käsebier, Clarence H. White, and Alfred Stieglitz; these photographers joined such illustrators as Jessie Willcox Smith and Elizabeth Shippen Green in depicting childhood as an idyllic, idealized realm apart from adult activity—a place where learning was natural and joyful. Magazines and advertisers used sentimental illustrations by Smith, Green, and their contemporaries to sell consumer goods; images of children sitting in window seats, painting with watercolors at kitchen tables, and cheerfully submitting to face washings promoted a sweet vision of middle-class domestic life to readers and consumers.[41] In these images, the innocent, adorable, and anonymous child was classed as privileged by virtue of being represented in distinctly upper-middle-class settings. These settings also tended to be—in direct contrast to Jacob Riis's alleyways, or Lewis Hine's factories—created especially for children.[42] Often, children in sentimental illustration and pictorialist photographs were pictured at play or reading books, representations that abetted the growing perception that a child was always learning and that learning was an idyllic process, carried out independently, but always with material support provided by a prosperous family situation.

In publishing photographs of museum children, taken by photographers whose names were not recorded, the *Children's Museum News* appended adorable captions, often using wordplay to gently humorous effect. These captions frequently emphasized to the adult onlooker the similarity between the natural objects that the children were engaging and the children themselves. The museum kept a hive of bees indoors; in a photo captioned "Busy Bees," a hive of children on their tiptoes gathers the "honey" of knowledge while observing the insects' activity. An image captioned "Opening Buds in Brower Park" depicts a group of boys and girls who may be observing botanical signs of spring, but the adult looking at the photograph is invited to think of the children as the buds prepared to flower. Considering the fear adults voiced throughout the 1920s about the prospect of working-class adolescent boys and girls mixing freely in social settings, this photograph is all the more notable for its presumptions of innocence.[43] In a photo captioned "What Beauties Heaven and

Nature Can Create," children look under a log for interesting animals and fungi, while adults looking at this picture are invited to see children anew as "beauties."

These "museum children" in the photographs were comporting themselves well, in accordance with what was expected of them. Child-raising advice published in the late nineteenth and early twentieth centuries emphasized the idea that children should be cheerful—that they should be obedient to parents' wishes, without sulking or strife.[44] Likewise, it was not just that the kids at the Children's Museum did not misbehave; they also actively enjoyed their time in the museum. (As the line in Anne Lloyd's poem went, they were "eager children" who made a sound like "happy humming," not only quiet children who caused no trouble.) According to its official accounts, child attendees naturally arrived at a satisfactory mode of behavior because they were so interested in the museum's activities. In 1913, the *News* touted its visitors' abilities to wait, writing that the small size of the lecture room at the museum meant that children might stand on line waiting for up to two hours on a crowded holiday. "The patience and good behavior of these youngsters is a constant source of wonder with us," wrote the anonymous author. The reason given for this "good behavior" was that many of the young visitors felt "proprietary interest" and had come "for definite purposes."[45] In a photo captioned "On the Line," small signs of the patrons' personalities, such as a look off into the distance, a face made, or a hat fiddled with, render the obedient waiting all the more notable—the implied message is that a child suppressing naughty impulses in order to gain access to knowledge is both adorable and admirable. In another captioned "Coming Early," museum patrons look like pilgrims, waiting for access to "their" house; the loose informality of the line, as well as the fact that this is Saturday morning, suggests that these kids are here on their own initiative.

One final image of children waiting to enter the museum serves to highlight the difference between what it means to look at children waiting out of interest versus children waiting because they must. In the Progressive Era, the emotions mobilized by images of children—whether learning or at play—varied widely according to the perceived relationship between the viewer's social position and the child's. Photographers like Riis and Hine used images of working children to shock and move audiences, dramatizing not only material deprivation but also the loss of educational opportunity inherent in childhoods spent laboring.[46] There is a big visual and emotional gap between the social activists' photographs of child workers and the serene portraits of chil-

Children "coming early" to the Brooklyn Children's Museum.
Children's Museum News, April 1926.

dren made by their pictorialist contemporaries. The working-class child, who is sometimes a child of color, appears exploited and in need of regulation; the middle-class child, who is always white, is idealized, abstracted, and observed with attention to nostalgic detail.[47]

In 1920, the Brooklyn Children's Museum began an "Americanization" program, which was mostly carried out in the American History section of the museum and consisted of discussions of American democratic processes and famous figures of United States history.[48] Programs of this sort were not uncommon at institutions serving children during this era; Americanization, reformers assumed, would be easiest taught to children, who would then impart the lessons to their parents.[49] A photograph of forty boys arriving at the museum to attend the program ran on the front of the *Children's Museum News* in March 1920. The boys in a double queue look miserable and cold in the "slush and fog," and one imagines them being forced to wait outside while the photographer gets the shot; possibly, the museum wanted to record this moment in order to thank the Red Cross for its help with transportation. These children, who were at the museum for very specific reasons (not to follow their own curiosity), are represented as safely orderly. They are not yet cute. Perhaps, after "Americanization," they will become so.

INVESTIGATIONS WITH PURPOSE

The Brooklyn Children's Museum tried to be quite clear about exactly what was learned at the museum and by which children. Museum tests were carried out to ascertain the amount of knowledge gained by individual students, and the *Children's Museum News* described these examinations as a pleasure, not a trial. Whether that was actually the case is, of course, less clear.[50] Tests were seen as being in line with children's own desires: in April 1916, for example, the *News*'s Library Notes reported that a "small boy" had picked out several mineral samples for the beginning of his studies in mineralogy, adding that this was "another step toward his expressed ambition 'to know all about everything in the Museum.'"[51] The bird clubs resulted in a visible increase in knowledge; in December 1913, the *News* reported that four boys who had become interested in birds the year before, Edward Crane, Carl Funaro, George Schoonhoven, and Wilfred Kihn, now had familiarity with a hundred living birds, although "Carl says he knew but three a year ago, and Edward thinks his own list could not have exceeded half a dozen."[52] Describing summer insect-collecting activities, the *News* wrote in 1915 that "no one was satisfied" until they learned the names of the species obtained.

Children transferred knowledge of the bird specimens in the museum into everyday life and schoolwork: "The effect of studying birds at home is registered in the children's daily conversations about living birds that they are beginning to observe in the parks and highways; in the discriminating questions that the children ask in the Museum, and the ease with which they can answer the questions asked in the bird lessons at school."[53] The strong connections between the museum and the Woodcraft and Boy Scouts organizations can be seen in the museum's stated mission to help Scouts and Woodcrafters study for badge tests.[54] Children competed for prizes in subject areas, striving to win such items as a "hand lens" (a prize for excellence in "insect study"); a book on trees ("tree study"); and a "balanced aquarium" ("the study of aquatic life").[55]

The field trip was a major aspect of the Museum's activities in the first quarter of the twentieth century, as well as a way for adults to teach children the habits of an ambitious and directed mind. The museum hoped to inspire children to look for birds and insects local to Brooklyn, and Anna Billings Gallup cited as successes several instances of children who became outdoor bird-watchers after their museum experience, using equipment lent to them by the museum.[56] Lectures given emphasized Brooklyn flora and fauna; lecture options for May 1915 included "Birds That Arrive in Brooklyn in May" and "Fur-Bearing Animals in the Woods Near Brooklyn."[57] In service of another league contest, students in mineralogy took "mineral collecting trips" into "nearby vacant lots."[58] Many of the field trips led by the museum during the summer of 1919 were to local spots, including the nearby excavation sites for the subway on Eastern Parkway, for mineralogy; Prospect Park, for bird-watching; and the shore at the end of Flatbush Avenue, for observing ocean life.[59]

Field trips, the Museum argued, helped teach children how to collect out of interest, rather than greed. Museum educators believed that children's enthusiasm for natural objects, when undirected, too closely resembled a mindless search for mass-cultural novelty; museum activities tried to channel this enthusiasm, remodeling the grabby habits of collecting children into a more disciplined program of acquisition.[60] In 1917, the *Children's Museum News* wrote about the progress of the summer insect-collection program, noting that the children had initially started out "with no purpose beyond that of catching and holding in their hands some brightly-colored bit of insect life" but that as they realized that the "cabbage butterflies, swallowtails with broken wings, and monarchs with wings rubbed colorless by too-eager fingers" did not present well when mounted, they moved on to more specifically directed collecting. In this scenario, participating in scientific modes of classification and presentation made the children abandon their desire to "catch everything that flies" in favor of a "businesslike purpose" that would curb acquisitiveness and result in an enhanced ability to render leisure time productive.[61] In another description of summer trips, the *News* wrote that children prepared their own collecting apparatus, saying that "many moments of happy anticipation" went into the exercise of scrubbing and varnishing cigar boxes, refurbishing butterfly nets, and readying cyanide bottles. (The *News* stretched perhaps a bit too much when reporting that the children also "appreciated the advantage of making their outfits neat and uniform in appearance, and keeping them in good condition.") Good behavior continued on the trip itself, as the children conducted their business without, according to the instructor accompanying the trip, "a word of friction, or a jar." "Placed in a field with

Patron George Ris identifying the minerals in his collection at the
Brooklyn Children's Museum. *Children's Museum News*, November 1917.

something definite to do such as insect collection or plant study the children
expand freely," the article continued. Appealing to the adult sense of the pic-
turesque, the author indulged in a moment of description: "In the open country, the children present an attractive sight flashing their white nets, chasing
butterflies, collecting wild flowers for the plant presses, and making vigorous
use of the geological hammers in search of crystals."[62]

The *Children's Museum News* often ran photographs of single children
working diligently on collections they compiled on these trips. One such was
a photograph of Grinnell Booth, a participant in one of the museum's summer entomology programs, in a quiet corner of the Adams House, processing
the fruits of his insect collecting. The accompanying text noted that Grinnell
was one of the most enthusiastic students in the program, "hesitating at times
about going home for lunch for fear he would 'miss something.'" (Grinnell was
also mentioned in the following month's *News*, as the youngest competitor in
an insect-collection contest, with an entry that "compared quite favorably"
with that of the older boys.)[63] These pictures of children working were clearly
set up for adult appreciation, as in a 1917 image of George Ris and his mineral
collection, in which the "trays of home manufacture" that house his specimens are turned outward for the camera to capture, creating the impression

of a little shopkeeper displaying his goods.[64] In her own photo opportunity, Bernice G. Schubert "of 1483 Union Street" was also shown with her collection (hers displayed in a Chiclet box) and described as "one of the happiest little girls in the Bedford Section all last summer"—a description belied by her somewhat doleful expression.[65] Next year, the *News* noted, she planned to raise butterflies from the caterpillar stage. These three images, grouped together, present a picture of children learning to direct inquiry properly. The museum itself acts as a stage for their quiet, well-behaved labors.

At the Brooklyn Children's Museum, a child's interest was expected to manifest itself in activities that would reciprocally contribute to the museum. In 1917, the *Children's Museum News* reported on several gifts children had given to the library, adding "the interest felt by the children in the Library has been expressed from time to time by gifts of books and useful material." (In this case, "useful material" included one Morton Wadsworth's gift, "an unusual Chinese nut, resembling the bronzed head of a deer, which has aroused the curiosity of children in visiting classes.")[66] One child's interest needed to multiply; such multiplication was profitable for both the museum and science as a whole. To receive a museum certificate in bird study, a child needed to "show" in what ways she had succeeded in interesting others in birds.[67] Children also assisted in museum promotion, as with the Children's Museum League begun in 1915, whose members were to wear a special badge and promote museum membership among their friends.[68] By 1920, the members of the league were enjoined to "make short addresses about the museum in classrooms or school assemblies" or "invite friends to bring box lunches and spend a whole day at the museum under your guidance." ("One member brought his friends on a bicycle one at a time from a distance of two miles.") For their efforts, league members successful at recruitment (having brought fifteen members into the museum fold) were given a bird book with "colored pictures."[69]

The museum's expectations of school-age children show how the utopian vision of an interest-driven kids' paradise was wrapped up in middle-class values of the time: community building, carefulness, thrift.[70] Most patrons were of elementary school age. In their teenage years, the patrons who continued to come to the museum took up pastimes that dramatized their new connections to a larger, more exciting world of scientific inquiry. In the record of wireless boys' museum projects, the contradictions between starry-eyed adult perceptions of children's hobbies and the often much more pragmatic investments by children in their interests can occasionally be glimpsed.[71]

BOYS' QUESTIONS, BOYS' FUTURES

"They argue almost to the point of the bayonet," Anna Billings Gallup told a reporter in 1912, describing the discussions which she saw take place between older, "earnest" boy visitors engaged in operating the museum's wireless station. The reporter, Sydney Reid, went on: "These heated arguments are not about baseball, football, tops, marbles, or kites. They are about moot questions of science. The big boys have exhausted text books, know all that the masters can tell about particular subjects, and are pushing their theories and inquiries into the unknown." Reid wrote that a group of older boys at the museum was so enamored of the wireless station that they "labor afternoons, Saturdays, and holidays, from love of their occupations, and because they 'want to know.'"

The maleness of this group—the boys' status as "big brothers" of the museum "family"—was a major part of its appeal. In her profile, Reid wrote, "Schoolgirls use the museum and its library freely, but they do no original work."[72] An article in the Brooklyn *Junior Eagle* in 1914 did mention two sisters—Katie and Fannie Weitzer, of 974 St. Mark's Place—who involved themselves in the wireless station, but their names never pop up in the *Children's Museum News*, and their futures as wireless operators, if they did continue with this interest, are not mentioned.[73] Within the museum's official publicity, out of all the activities carried out in the Brooklyn Children's Museum, the most glamorous club of all—the club that did scientific work that involved climbing about on roofs and performing "original" theorizing—was one made up only of boys.[74]

The romance of museum "big boys" was part of a larger cultural affection for the young inventor-hero, who, during the years between 1906 and the U.S. entry into World War I in 1917, was often adept at wireless communication. The subculture of boys and young men who constructed and operated amateur wireless stations was accompanied by—or perhaps cocreated with?—strong media interest in the phenomenon. Newspapers covered wireless operators with breathless praise, as in the 1909 case of Jack Binns, a twenty-six-year-old wireless operator who saved a group of incoming Italian immigrants from shipwreck when he spent hours sending a distress call to nearby vessels. Newspaper reports often emphasized Binns's youth when recounting his heroic actions. Children's books and magazines joined in the lionization of wireless operators, and many high schools had wireless clubs, while importers and marketers made basic equipment available at a price that rendered it accessible to middle-class children.[75] Meanwhile, the contemporaneous nature

High school boys on the roof of the Brooklyn Children's Museum, attaching wires to a pole. *Popular Science Monthly*, April 1908.

study movement did not produce textbooks and curricula that covered the physical sciences. A combination of factors, including the greater involvement of women (and, thus, of future teachers) in biological sciences, as well as the cultural belief that young children had an affinity for animals and plants, led to this state of affairs.[76] The Children's Museum library seemed to follow this belief, as it considered books about physics and astronomy to be appropriate for older museum patrons, while describing nature study books as intended for their younger "museum children."[77]

The emergence of the wireless station at the museum seems to have resulted from children's desires. Gallup said that the museum put together a lecture series in physics in 1906 "in response to an expressed demand from the boys." This course led to the initiation of the wireless station project, which Gallup said was exclusively run by the "boys" themselves. Gallup's language indicated that this innovation on the Children's Museum's part was not received entirely positively in the greater community of museum workers; she said that "some have maintained that physics and electricity are not germane to museum work," which should remain focused instead on collecting and cataloging objects of scientific interest, but argued that "a children's museum calls for such modifications and adaptations of methods as will enable children to use it" and that "the keynote of childhood and youth is action."[78] An assistant curator, Mary Day Lee, who began working at the museum in 1907, became a de facto specialist in these "older boys," offering lectures on minerals and physics and helping the boys with their wireless station.[79] The place of these "boys" in the museum family was a treasured one. Reid reported that, in return for the space that Lee and Gallup afforded them to work on the wireless station, "if a fuse blows out or anything goes wrong with museum apparatus, the ingenious and industrious boys immediately fix it . . . . Led by James Parker, they installed a complete telephone service for the museum, and this has worked well during three years."[80]

Although some wireless amateurs challenged authorities by clogging airwaves with their communications, the boys at the museum carried out their wireless experiment within the parameters of the establishment. In 1915, the *Children's Museum News* reported that the boys of the wireless station were working hard to meet government requirements regarding wavelength and dampening levels. When the United States entered World War I, the wireless station was shut down, by order of the government; the *News* reported with pride that many boys previously active at the museum's wireless station were now serving in the military.[81]

Museum staff reported that the wireless station produced successes, as did many of the museum's activities, by directing children's interest into a productive context. In 1915, the *Children's Museum News* published an account of a mother who came to the museum with a younger child, and when the child introduced her to a staff member, she launched into an account of how the museum had saved one of her older boys from "incorrigibility." This mother's story offers a window into the way that the group of boys working in the wireless station took science as the foundation for their community. Before she moved to the museum's neighborhood, the parent said, she had despaired for her son's future; "the neighbors advised me to thrash him." However, after visiting the museum, "he awoke to an interesting world." The museum removed the obstacles to interest that were artificially placed by the school system between her son and the world, and the child became devoted to wireless telegraphy. In the wireless station of the museum, "he handled apparatus, asked questions, and performed experiments." His peers helped him see the value of scientific discourse: "He heard the discussions and arguments of the older boys who frequently disagreed on scientific questions. . . . As a listener, and sometimes as a participant in these heated arguments, he learned to do his own thinking." As time went on, "the Museum became his play room, his study, and his work shop all in one," while he developed new ambitions and ended up enrolling in a technical high school and pursuing a degree in electrical engineering. The museum, the happy mother continued, "applied the stimulus and continued the encouragement until the boy was old enough to make his decision and plan his own future."[82] This redemption story showed how appropriate boys' energies were for scientific pursuits; the account of the magic turn from destructive to constructive would be echoed many times in stories told about boys' scientific vocations in the twentieth century.

The "wireless boys" offered one of the museum's best opportunities to prove that "its" children were long-term successes. In 1916, for example, the *Children's Museum News* recorded that the world, along with museum staff, had recently been excited by announcements of wireless communications between Arlington, Virginia, Honolulu, and Paris: "Such an achievement as this stirs the imagination of all people, but to the staff of the Children's Museum there was added a deep personal interest." Two young men, Austen Curtis and Lloyd Espenschied, who had been instrumental in the initial opening of the wireless station at the museum, were involved with these tests. In the accompanying photograph, taken while the boys were still living in Brooklyn and attending the museum, Espenschied looks somber, serious, and focused; Curtis leans away from Mary Day Lee, looking raffish, almost mischievous. The article notes that while Espenschied had graduated from a technical school and worked for AT&T, Curtis had learned in the "school of experience," traveling to "many lands" as a wireless engineer, living in Brazil and finally returning to work for the Western Electric Company.[83] An article in the *Brooklyn Daily Times* pointed out that Curtis, a "particularly efficient" young man, was "the chief wireless engineer for the Brazilian government before he was old enough to vote."[84] When he came back from Brazil, Curtis brought the Children's Museum a collection of tropical insects and, later, a live spider monkey named Plato, who was to become a favorite museum pet.[85] During the First World War, the *News* reported that Curtis, a First Lieutenant in the Reserve Signal Corps, was "engaged in very important specialized wireless work."[86] In 1918, Curtis wrote to the *News* to announce that he had been promoted to captain in the Radio Corps, and in 1919, he came back, along with a number of other museum "alums," to visit the curators and staff.[87]

The war, in fact, provided many opportunities for the *News* to catch up with the wireless alumni. In 1918, the *News* reprinted in its entirety a letter sent by a soldier who learned radio at the museum. In jaunty prose, the letter writer professed best wishes to all the boys at the station and called himself "a small part of the museum contribution toward the war." "Perhaps the big bugs that pull the strings didn't do such a bad thing after all when they gave the boys a few instruments to play with," he concluded.[88] A letter from another, unidentified wireless graduate, excerpted in another issue of the *News*, showed how modernist ideology, founded on technology, went hand in hand with patriotic sentiment: "Everywhere the American Army goes we clean up, for our modern ideas and thoughts keep us from drifting into century-old ruts. I think our modern ways will awaken the Europeans and give them a different light."[89]

The wireless boys were the best example of follow-through received by the museum from its "graduates," but other examples of successful "alumni" abounded. The Children's Museum could, and did, point to children whose experience with the museum led to a career in science. At the National Education Association in 1926, Anna Billings Gallup gave a speech that highlighted the usefulness of these former patrons to society. The museum's displays, she said, "fired one boy with a zeal for insect lore that wrought his way through the University and eventually saved from a threatening insect plague, the wheat crop of Indiana." The mineral room "[inspired] one boy to become a curator of minerals and two to qualify as mining engineers."[90] Along the same lines, in 1929 the *News* reported on museum "alumni" Foster H. Benjamin, who had become an entomologist and was currently a "technical expert" in the "battle against the Mediterranean Fruit Fly," which was threatening Florida's fruit crop.[91] Gallup's rundown included no women; if the museum prided itself on taking boys from the "chicken, chicken" stage to the acquisition of useful technical expertise, there was not an expectation that the girls would follow suit. The prized place of the wireless station in the museum's life was a reminder that the best, most interesting scientific work was reserved for boys.

CONCLUSION

It's sweet to look upon boys or girls with butterfly nets; it's thrilling to see teenage boys on the roof, hooking up systems that their elders do not understand. In their representations of what went on in their institution in the early twentieth century, the educators of the Brooklyn Children's Museum painted a picture of a thriving kids' culture with science at its heart. They also shaped their own vision of what it might mean to be a "modern" person: a cheerful, Americanized, curious, white child, capable of far more than his parents might imagine. Acquisitive, sexual, or disobedient impulses would vanish in the face of the charms of the museum—and of science.

By shaping the museum as a quasi-public space exclusive to childhood, the museum's project solidified the idea that children (at least children coming from roughly the same social class) had a lot in common and that one such common factor was an active interest in science and nature. Adults looking on could picture visitors to the museum as participants in a generative utopia, where modernity—in the shape of the Children's Museum idea—worked to offer just the right amount of satisfaction for children's curiosities. If everything else about the museum was natural (or so the implicit argument went), then the slow separation of the older boys into professionally oriented, technologically savvy groups must be too.

Science in the Basement

Selling the Home Lab in the Interwar Years

The cover of a kids' biography of Thomas Edison, published in 1933, featured a red-cheeked and happy young Thomas Alva. This "Al," as author Winifred Esther Wise calls him, is in the mobile chemical laboratory he set up for himself on a train car of the Grand Trunk Railway, where he worked selling food to passengers. The cover designer, Charles Ropp, took pains to showcase all the evidence of Edison's famous industriousness: the picnic basket full of wares, the row of chemicals in stoppered bottles on the wall, and the pile of newspapers, which, as the book told young readers, Al had edited and printed himself. Al's happy face, in alarming proximity to the squiggly lines indicating upward motion from the vessel into which he has poured his chemicals, is a vision of experimental ecstasy. Edison, who died in 1931, was the subject of a spate of hagiographic biographies and biopics in the 1930s; he had been a boys' book favorite for decades, and authors loved to point readers to his entrepreneurial spirit, energy, and perseverance (a 1916 review of an Edison biography in the Boy Scout publication *Boys' Life* began: "Talk about your live wires, Edison was surely that.").[1]

Wise, like boy-book biographers before her, was clear: "Al" was no oddball, despite his solitary pursuits. He was a little (maybe a lot) better at experiments than other boys, but he "could chew tobacco, spit as far as any of the fellows, and think up as much mischief."[2] This depiction of Edison's other

The cover of Winifred Esther Wise's *Young Edison: The True Story of Thomas Edison's Boyhood* (1933). Rand McNally.

activities highlights the didactic message of such biographies: An interest in science was a shared boyhood hobby, normal, common, and communal. In a memoir of his boyhood tinkering in the 1930s, historian John Lienhard describes, with some nostalgia, what he characterizes as a nationwide culture of boyhood experimentation: "We all felt the thrill of knowledge passing through our fingers and into our brains. Curiosity grew with each failure, and confidence grew with each success."[3] Lienhard, along with other boys and adults living at the time, perceived boyish laboratory activity as part of the zeitgeist.

The strenuous boyishness of this culture of experimentation replaced a previous culture of home science that left more room for the whole family to participate. In the nineteenth-century United States, a middle-class ideal of family life made room for scientific demonstrations in the parlor or advocated specimen collecting as a group activity. Books for children, such as the Rollo series by Jacob Abbott or the many children's books on astronomy or chemistry published in Britain, depicted scientific entertainments that families might try together, using household objects; they often pitched these entertainments to both boys and girls.[4] By the interwar period, home chemical experimentation became the cultural province of American boys, who took over corners of the now much-more-informal family home to install their laboratories. While the home itself underwent a technological revolution, becoming increasingly "modern" in its layout and appliances, and mothers and wives were called upon to approach housekeeping as a science, it was the boy in the basement with his chemistry set who was hailed as the closest thing the family had to a direct connection to the world of science and industry. In his individual fumblings, adults saw the grand possibilities of the future.

The image of the nation's many Edisons—industrious boys assiduously cooking up mischief in some unsupervised basement, attic, closet, or treehouse—is still a dearly held one. Many people commenting on present-day cultures of childhood offer reminiscences like Lienhard's, harking back to what they see as a lost scene of youthful experimentation, one that has become an iconic American image. This particular icon is hardy, charming, and difficult to question. Yet, if we turn the twentieth-century boy chemist over and shake him a bit, questions fall out of his pockets, along with the snips and snails and puppy dog tails: Was he ever truly real? Why do we love him so much? And where is his sister?

A CULTURE OF FACTS

During the interwar years, middle-class Americans encountered discussions of scientific activity increasingly frequently in their everyday reading, and the authority of "science" as a cultural force grew.[5] Even as some objected on religious grounds to evolutionary theory, raised concerns that experimentation on live animals or vulnerable human populations was inhumane, or protested the use of chemical weapons in World War I, the pursuit of science was increasingly allied with other national goals and acquired a newly potent ideological force.[6] Scientists tried to convince the public that coordinated research, sponsored by the government, was necessary and worth investing money to support. The experience of World War I, during which it became clear that American industry's use of science was far behind that of the more sophisticated Germans, served as a spur and a wake-up call.[7]

If some interwar science promotion was informed by fear and by desires for national dominance, science boosters in these years also offered an enticing vision of bounty, pointing to the material comforts that scientifically informed industries could provide. Writers of science- and technology-focused books and articles in the interwar years related science closely to the achievements of industry—and, in turn, to the beneficial changes science had wrought in everyday life. Popular writing about science for adults emphasized the magical, life-altering qualities of the new substances coming out of laboratories, striving to evoke a sense of the marvelous. Paul de Kruif's *The Microbe Hunters* (1926) and Sinclair Lewis's *Arrowsmith* (which won the 1926 Pulitzer Prize) celebrated the achievements of medical researchers.[8] Big cultural commemorations, such as the 1933 Century of Progress International Exposition in Chicago and the 1939 New York World's Fair, showcased newly available technological wonders and offered visitors a vision of future abundance. The 1933 fair's motto was "Science Finds, Industry Applies, Man Conforms"—a top-down vision of the value of new knowledge that illustrates how much cultural power science had accrued.[9]

If science and industry were all-powerful, the average person had correspondingly more avenues for keeping mental pace with modern life. The parents of the boy with the chemistry set were increasingly aware that American middle-class culture believed in the importance of knowing how the world worked. Titles intended to introduce readers quickly to complex subjects in history, philosophy, and science were some of the most popular of the 1920s, with authors like H. G. Wells, Will Durant, and Hendrik van Loon seeing lucrative returns on "outline" books that promised complete knowledge of a topic in minimal reading time.[10] The publishing firm Simon and Schuster began awarding their Francis Bacon award at this time, named after the great empiricist "in recognition of his own daring and monumental achievement in taking all knowledge for his province": "The purpose of the award is to stimulate and reward the writing of books which . . . 'carry on the conscious adventure of humanizing knowledge.'" Judges for the award included a trio of nonfiction luminaries: Durant, van Loon and Edwin Slosson, of the nonprofit science journalism advocacy group Science Service.[11] John Dewey and astronomer and science popularizer Harlow Shapley sat on a council advising the jury. "Effectively organized and animated, truth is infinitely more romantic and more exciting than fiction," the sponsor of the award wrote.[12] The pursuit of truth, new cultural institutions like Science Service and the Book-of-the-Month Club implicitly and explicitly argued, was proper for middle-class Americans. Whether or not such readers really wanted to "find out" about the new laboratory-based medicine, the Roman Empire, or the history of invention, the *appearance* of curiosity was socially encouraged.

One population — children — had the prized modern quality of curiosity in abundance. The idea that children naturally sought facts about the world was key to the philosophies of progressive educators. Maria Montessori, writing of the observations of playing preschoolers that shaped her approach to early childhood education at the turn of the twentieth century, recorded a realization: "As I stood in meditation among the eager children, the discovery that it was knowledge they loved, and not the silly game, filled me with wonder and made me think of the greatness of the human soul!"[13] This seductive epiphany, which painted humans as fundamentally knowledge-seeking beings, identified children as the purest knowledge seekers of all — a belief that both flattered childhood and provoked a new set of prescriptions for the kinds of things children should be doing in their leisure time.

As book acquisition and reading established themselves as part of the middle-class child's life, with children's rooms newly accessible at libraries and a growing children's publishing industry enforcing the idea that children should spend a lot of time with books, the "informational" book and encyclopedia was central to that child's shelf.[14] In a 1930 article in *Parents* magazine, author and journalist Maude Dutton Lynch wrote that if parents would simply ask children what they wanted to read, they would be surprised by their offspring's need for information about the "real" world. Children, Lynch and other experts held, were motivated to read not to enter a reverie and lose sight of the world but to bring that world into closer focus. "[Children] want to know the story of mankind and the story of the earth," Lynch wrote. "They want to know all about ships, and railroads, and aviation; they want to know how books are made and the story of milk. They want books about electricity, about the stars, about moths, about how cities build their water supplies. They want encyclopedias of their own to turn to in time of need and to roam about in, urged on by the itch for more knowledge." Publishing companies, happy to oblige, had begun to put out "books to which the child could turn himself to find the answer to his many 'hows,' or to read along the line of his particular interests at any particular time." Providing these "information" books could help children learn to read quicker, Lynch went on, arguing that interest in the big world would provide an indispensable stimulus to continued learning.[15]

The researching child, lost in the pleasure of "finding out" facts, represented the potential for a harmonious individual relationship with the proliferating knowledge available to modern citizens. In fact, those editors and authors within the children's book publishing industry who advanced this argument said, children were great at wanting to know about all the new things around them — maybe even better than adults. Commemorating the careers of husband-and-wife author-and-illustrator team Maud and Miska Petersham in the *Horn Book Magazine* in 1946, on the occasion of their winning a Caldecott Medal, Irene Smith Green recalled an incident that proved the popularity of the Petershams' illustrated subject books about the histories of various commodities: "I can remember a little girl on Christmas Day in 1939 running about the house, neglecting toys and picture books, hugging *The Story Book of Rayon* and caroling to her parents' astonishment, 'Goody for good old rayon!'"[16] The anecdote combined wonder at this eminently empirical childish interest with the evocation of a very new material, rayon — a fabric that would have been introduced during these parents' lifetimes and had been named only in 1924. That the child should consider a *modern* fabric an object of fascination particularly pleased the onlooking adults.

Children's desire to read nonfiction, adults in the publishing industry argued, was an outgrowth of the changes of modernity, which had altered the very nature of their interests. In 1919, Franklin Hoyt, a publisher in Houghton Mifflin's department of education, made a speech at the meeting of the American Library Association in which he argued that modern life rendered fiction completely unattractive to children; they should, instead, have stories about the world itself.[17] "It is only the blind eye of an adult that finds the familiar uninteresting," progressive educator Lucy Sprague Mitchell wrote in the introduction to her influential *Here and Now Story Book* (1921). Stories should "further the growth of a sense of reality, give the child the sense of relationship between facts, material and social: that is to further scientific conceptions."[18] Children's interest in the relationship between these facts would naturally provoke "inquiries which hold the germ of physical science."[19] At least some children's book editors agreed with this assessment and came down on the side of "milk bottles" rather than "Grimm" (as Virginia Haviland of the Library of Congress defined the divide);[20] in 1930, Scribner's editor Alice Dalgliesh wrote in *Publishers' Weekly* that "modern life is full of interesting, real things, and there is no time for sugary little fairy tales of the type that used to be published by the dozen."[21]

The children's books of the 1910s, 1920s, and 1930s featured many titles that began with "The Story of" and ended with the name of an important commodity. This naming convention could also be found in youth-oriented encyclopedias — Henry Chase Hill's *Wonder Book of Knowledge*, for example, was composed entirely of "stories of" sandwiched between sections of questions and answers. Lucy Sprague Mitchell contended that these kinds of stories of real life could appeal just as strongly as "tales of hunting, of impossible heroisms, and of war." "The world of industry," she wrote, "holds possibilities for adventure as thrilling as the world of high-colored romance."[22] The convention of including the word "story" (or, occasionally, "wonder," as in "The Wonder of Oil") in the titles of these books reflected the belief that children would find narrative and excitement in an expository explanation of industrial processes. At the same time, this interest would yield real knowledge. Mitchell's own life provided a model for how to incorporate education into family time. Around the time the British children's encyclopedia *Book of Knowledge* became a fixture in American childhood libraries, Mitchell's husband read to their son Jack from the *Book* every night; afterward, Jack was required to prove what he learned by dictating an abstract to Lucy.[23]

The prescriptions of children's book authors echoed beliefs held by educators trying to teach "modern" interwar children, thought to be naturally attuned to the ever more elaborate human-built structures and systems that shaped their world. The general-science curriculum, which educators conceived as a more appealing way to introduce younger high school students to science than specialized discipline-based classes, operated on the basic principle that the students should be introduced to science by learning the underpinnings of their technological world. Like the "story of" books, general-science classes linked industry and the domestic, skirting difficult terminology in favor of helping students make connections between scientific principles and everyday life.[24] This was also a principle of progressive educationists working in the younger grades, who worked backward from familiar objects and processes to explain the web of technologies that connected the world.

These approaches to education presumed that children's interest in the details of the "wonderful" processes of industry was nothing less than insatiable. This presumption reinforced ideas about both childhood and industry; in these books and in classes in general science, children were educated in the importance of being curious about the mundane and the procedural, while also learning that American industry was all-powerful, morally irreproachable, and increasingly rational in its practices. Picture books about such spectacular technologies as trains and skyscrapers and ABC books featuring "modern" objects were replacements for the "fairy tales" disdained by progressive educators seeking reformations of children's literature.[25] But these nonfiction books about industry were another kind of "fairy tale," reassuring adult authors, editors, and parents about children's capacity to understand the increasingly complex processes of the modern world.

CHEMISTRY, THE BIG AND THE SMALL

It was in the context of this self-consciously modern culture of facts that the children's chemistry set found its market. Beginning in 1918, the A. C. Gilbert Company of New Haven, Connecticut, the Porter Chemical Company of Hagerstown, Maryland, the Lionel Company of New York City, and various smaller companies produced, marketed, and sold a wide array of science sets meant to introduce children to topics in chemistry, microscopy, biology, electricity, and physics. Technological hobbies likewise expanded, with sales of parts, instruction booklets, and kits offering children and young people the chance to pursue interests in wireless, ham radio, aviation, and automobile design.[26] The science set joined the encyclopedia, the industrial storybook, the model airplane, and the wireless set in the pantheon of home equipment suitable for modern children whose play (as Montessori put it) was their "work."[27]

The marketing of these sets, their packaging, and their textbook-length manuals all respond to popular interwar assumptions about the manly nature of science, perceived as a military, industrial, and entrepreneurial pursuit. Adults imagined that the social world revolving around the chemistry set would be an almost exclusively male one, arguing that sets would promote confidence and mastery. Such men as A. C. Gilbert, a physician, former

Olympian, and patriotic devotee of such pursuits as hunting and shooting, leveraged their own masculine identities—forged in a Progressive Era culture strongly influenced by Theodore Roosevelt—in order to sell science to boys. Chemistry sets emphasized the practice of science as a pursuit intended to sharpen natural curiosity and, at the same time, strengthen children's more worldly powers of observation, self-regulation, and entrepreneurship. It is this era, which presumed that American boys would pursue chemistry without adults and with gusto, that has shaped today's nostalgic vision of the naughty, inquisitive "boy with the chemistry set."

The growing public profile of the chemical industry abetted chemistry-set manufacturers in their promotional efforts. In the late nineteenth and early twentieth centuries, the industry's profile in the United States grew. Around the turn of the century, DuPont, previously a gunpowder manufacturer, shifted to producing chemical explosives such as TNT, in the process transforming itself into a chemical company. Dow (1897), American Cyanamid (1907), and Union Carbide (1917) were all founded in the space of a few decades. During World War I, American chemical companies met the increased demand for chemicals alone, as trade with Germany temporarily ceased; the experience gave the American industry a boost and forced innovation in several areas. After the war, DuPont expanded, becoming one of the ten wealthiest American companies by 1929.[28]

The middle-class culture of facts embraced American chemistry. While some peace activists publicly questioned the science's involvement in the development of weapons during the Great War, voicing doubts about the morality of this new way of war, they were a minority voice in the interwar years. Industry groups undertook a targeted crusade in the years before and after the war to promote chemistry in the public mind, going so far as to claim chemical warfare as a more humane method of attack than conventional bombs and bullets.[29] These years saw the publication of several popular nonfiction books glorifying the industrial and medical contributions of chemistry (for example, Edwin Slosson's *Creative Chemistry* [1919]), and scientist Marie Curie enjoyed celebrity in the United States, touring the country to great acclaim in 1921.[30] Boosters presented chemistry to the public as a science that combined the appeal of pure intellectual engagement and potential for social good with strong possibilities for industrial exploitation and profit ("To Make Money: Use Chemistry," an article in the *Literary Digest* in 1927 was headlined).[31]

Marketers of chemistry sets for boys embraced these lines of argument,

The cover of the Porter Chemcraft Chemical Outfit No. 1, circa 1920s.
Courtesy of the Chemical Heritage Foundation.

hitching themselves to the industry's rising star. The artwork on the box tops of chemistry sets established a set of visual clichés—an iconic language that was meant to connect the child's work in the home lab with the grand project of industrial chemistry. The artwork on Porter's chemistry sets, whose motto during the pre–World War II years was "Experimenter Today, Scientist Tomorrow," often included images of adult scientists in white lab coats. On some boxes, such as that housing the Chemcraft Set No. 1, the boy scientist was framed by the twin images of the adult male scientist and a factory. The inclusion of this imagery in the material that met the prospective buyer's eyes invoked the invisible supervision and guidance of the adult scientist, in a way that also promised a future professional incarnation of the child who used the set.

Science toys were a miniaturization of an overwhelming force in the social world—science—into a physical package. Unpacked, science sets unfolded into small laboratories, staffed by children, who would in return produce more science for the world to use. Later in the 1920s and the early 1930s, such sets, which had been sold packaged flat, began to include racks that would allow the young experimenter to array the chemicals vertically or to be sold in cases that stood upright, displaying the chemicals in neat rows. Companies called this innovation "laboratory style"—another way in which the companies were selling the idea that children could use kits to emulate scientists in their labs.[32] Box tops' and manuals' inclusion of imagery pointing outward to the places outside the home, where chemistry was professionally pursued, emphasized that connection, selling the concept of a chemistry set to adults who considered themselves forward thinking and science minded. The idea was designed to be pleasing to adult buyers: to purchase a chemistry set was to abet the replication and expansion of modern knowledge, to bring the productivity of American industry into the home.

This advertising approach was important because it surmounted adults' misgivings about buying too many toys for their children. Some adults in this era found children's presence in the newly expanded consumer sphere vaguely upsetting, wondering whether what felt like an endless parade of novelties now available to children, together with advertisements hawking the latest in cheap manufactured amusements, would dampen their taste for more wholesome pursuits. Not all childhood consumption, however, was perceived as an equal moral hazard. In the 1920s and 1930s, such child development experts as Angelo Patri and Sidonie Gruenberg, who enjoyed a new prominence in magazines and books directed at parents, framed kids' involvement with the marketplace as neutral—and possibly, with proper guidance, beneficial. Children should be given allowances and taught how to use them. Boy consumers, especially, could channel consumptive desires into productive hobbies that might help them succeed in later life.[33] With this recasting of the meaning of youthful consumption, a boy who desired a chemistry or science set was not greedy but instead possessed a laudable interest in something universally respected: science. Meanwhile, a parent enabling a child's consumption of these sets was advancing a child's chances of embarking on a lucrative career—a goal that was of particular interest to the parents of the 1930s, who worried that their own diminished material circumstances might stunt their children's chances of success.[34] A parent enabling a child's consumption of these sets was advancing a child's job opportunities, while contributing to the nation's industrial advancement. Such a parent was even, as one writer argued in 1931, enhancing possibilities for world peace.[35]

Manufacturers of science sets and other educational toys self-consciously claimed positions as public servants, encouraging "good" consumption that would, in the long run, further national goals.[36] Marketers labeled anything from the widely used construction sets to play carpet sweepers for girls as "educational." In 1916, in an item headlined "Inspiration," the industry publication *Playthings* trumpeted: "What philosophy, science, and art are to civilization, business to man, the fireside to woman, toys are to youth. Toys are the child's WORLD!" When gazing at the toy, the anonymous author went on, the adult should realize that "here is Science at its source; Art in its adolescence; Power at its portal!"[37] This was an intoxicating premise for an industry looking to make a place for itself as a rational and legitimate arm of public culture. For the relatively young American toy industry, which was coming into its own during these decades, making educational toys was a good way to claim status as inherently "American." In 1917, a writer for *Playthings*, Thomas K. Black, mused about an editorial he had read in a newspaper which had claimed that European manufacturers tended to produce toys that "cater to the child with the whimsies, the fairyland notions of childhood," whereas American toys "quickly become practical, and in a way useful, as being educational and suggestive." Black embraced this assessment (while, of course, patriotically arguing that American toymakers also produced wonderful, whimsical playthings, which the author had overlooked).[38] By pointing to its homegrown manufacturers of educational toys, the American toy industry could differentiate itself from much better established European competitors and answer parental worries about consumption at the same time.

Of the major companies that produced chemistry sets, the A. C. Gilbert Company was the most intent upon claiming a place for itself as a national utility: a developer of boys, kindred to the many youth groups, camps, and programs that sprang up during the Progressive Era.[39] During World War I, A. C. Gilbert's ads in *Playthings* argued to fellow toy producers that while wartime necessarily meant sacrifices, people would still buy Gilbert. "Gilbert toys are essentials," the ad proclaimed. "Gilbert Toys are essential instruments of Americanism. The vision, the creativeness, the initiative that Gilbert Toys develop and encourage in a boy are the very qualities of Americanism that made our soldiers the wonder of Europe, that made the world gasp in amazement at the raising and transportation of our army."[40] Gilbert regularly deployed this identity to argue to readers of *Playthings* that they should stock Gilbert toys before all others. In early 1920, Gilbert posited that the recent postwar economic downturn meant that people were going to want to buy toys different from "the hanky-panky toy, which comes today and is gone tomorrow." "Gilbert Toys," the ad copy maintained, "don't just happen. They are built with the great motive of educating the tremendous army of boys throughout the world. They mean something. They are genuine."[41] By claiming the ability to develop "vision, creativeness, [and] initiative" in boy customers, Gilbert positioned himself at the wellspring of American power; this advertising strategy painted his toy company as a public good, rather than a profit-making enterprise.

In keeping with this image, the toy industry paid attention to new theories of child development—attention that assisted the industry in its selling tactics, while providing a veneer of wholesomeness and carefulness for toy companies. *Playthings* often published guides for "toymen" (as they called themselves) wanting to know which toys to sell to which kids.[42] Chemistry sets and microscopes, while sold in child-friendly kits, offered a maximal sense of "realism"—an effect that, everyone agreed, was what children of elementary school age wanted from their toys.[43] "It is a very familiar fact that if a grammar-school boy once gets into the real creative world of industry he can hardly be dragooned back to textbooks and school routine," *Playthings* wrote in 1919, showing that toymen were aware of educational theory and used it to their benefit. "The big thing in education is to so link up the school with the visible, bustling world as to keep the child's workman-like instincts engaged."[44] For their part, interwar experts who advised parents on child development included chemistry sets and other educational toys in their recommendations. Writing in 1934, child psychologist Ethel Kawin told parents trying to choose toys that children ages ten and eleven had a "tendency to give increasing attention to objective reality." As opposed to the younger child, who would be "concerned with things as he knows them from his own experience," Kawin wrote that the older child "begins to grasp the idea that things are not merely what they seem, and he develops a genuine desire to know them and to understand them as they are." These older children "are no longer satisfied to push a button and see things move. . . . They are interested in using and in understanding machines which transform energy; they are likely to show an interest in simple chemistry sets; they are fascinated by electric-train equipment and can now begin really to study and to understand the principles upon which such apparatus operates."[45] Supported by such pronouncements from specialists in child development, chemistry set manufacturers represented their toys as completely natural — and necessary — companions for growing modern children.

Just as the publishers of children's books during this time were convinced that nonfiction "information" books would excite children's interest more than would fairy tales, the toy industry marketed science sets using the presumption that children prized verisimilitude above all qualities. In a 1946 article, a journalist wrote of Gilbert's philosophy: "Kids demand the utmost in realism. At the Gilbert Hall of Science in New York . . . the men in charge never talk to young visitors about 'toys'; they talk of such things as 'structural steel engineering' and 'chemistry laboratories.'"[46] In 1941, Lionel, introducing its first chemistry set, emphasized the sturdiness and realness of the set's equipment in an advertisement in *Playthings*: "It's not a flimsy, good-for-a-few-days plaything, but a whole research laboratory in compact miniature." The advertisement shows the set spread out on a table, with each piece flagged with information about its authenticity ("genuine Wedgewood mortar and pestle"; "rigid, substantial test tube rack").[47] This approach complimented the taste and knowledge of the boy customer and his parents. The companies also advertised the involvement of actual scientists in the development of their sets. In 1920, Gilbert told consumers: "The outfits we now show are the results of more than a year's work by Mr. William J. Horn, Ph.D. of Yale University, who has concentrated all his efforts since joining our staff, to producing outfits of an intensely practical nature, with which some astonishing experiments can be performed."[48] The invocation of these scientific credentials reinforced the argument that Gilbert sets allowed for an authentic connection to the world of "adult" science.

As these appeals were directed at parents and at toymen who might stock Gilbert or Chemcraft toys, different arguments were made to prospective child buyers. A. C. Gilbert was particularly intent upon understanding (and, it must be said, normatively *defining*) the psychology of the boy who might buy his sets. Throughout the first half of the twentieth century, Gilbert's company followed trends in children's culture in its promotional efforts, allying itself with popular cultural forms while purporting to offer educational content. In the United Kingdom during the interwar years, the manufacturer of Construments kits, which allowed children to make their own scientific instruments, sold the toys as a stopgap meant to fill in holes in the education system.[49] In the United States, on the other hand, A. C. Gilbert thought association with formal schooling would be a disaster for his toys' appeal. "When schools became interested in our construction and educational toys," he told his biographer, "we discouraged them as much as we could.... We were afraid that if kids saw our things in school, they'd think they were just as deadly dull as the rest of school and would have nothing to do with them."[50] For Gilbert, science and engineering were best done outside of school; chemists could be "self-made" in their knowledge and did not need to pursue success in home labs in order to please teachers. In fact, the very idea of the teacher-student relationship was antithetical to the picture of free boyish experimentation that Gilbert was trying to sell.

In keeping with this philosophy that experimentation should only be associated with fun, Gilbert's "curriculum" tied science into dominant trends in children's popular culture. When radio serials were popular in the early 1930s, Gilbert sponsored and hosted a short-lived radio show: "Thrills of Tomorrow for Boys."[51] In 1941, the company opened a store near Madison Square Park in New York City, the Gilbert Hall of Science, which had a hybrid identity: store and museum. Gilbert was on hand at its official opening, which featured 1,500 boys as "guests"; he dedicated the hall to the boys of America."[52] In the 1950s, the company sponsored a television show broadcast from the New York Gilbert Hall of Science: "Boys' Railroad Club."[53] In the comic-happy postwar era, sets included Gilbert-produced comic books, illustrating uses of Gilbert equipment.[54]

The manuals included in chemistry and science sets, which were intended for boy users to read (without necessarily speaking to parents), are important indicators of the way that companies tried to "sell" boys on science activity. These manuals were like miniature textbooks: Gilbert's 1920 "Gilbert Chemistry for Boys" book was 217 pages long. In 1918, Gilbert advertised chemistry outfits including a manual "giving complete instructions in boy language";[55] despite this boast, the manuals, which packed in the history of the science, technical instruction in the running of a laboratory, and reporting on the current state of the chemical industry, are heavy fare. They were often produced, or at least edited, by scientists, who lent their credibility to the packaging. A 1936 Gilbert manual's title page featured not only the assertion that the manual had been produced "under the direction of the Bethwood Research Laboratory, Bethany, Connecticut," but also the names of two Yale Ph.D.'s, Treat Johnson and Elbert Shelton, as coeditors, along with Gilbert, whose M.D. from Yale was prominently featured.[56] A Gilbert ad that ran in *Boys' Life* in 1930 framed the manual as a direct communication between adult chemists and boys: "an *interesting* book explaining the chemist's secrets, with which boys can do the new things expert chemists have discovered."[57]

In the prose of the manuals, a picture emerges of the selling techniques that these companies used to convince children to be interested in chemistry and microscopy—and through these techniques, the opinions these companies held of the nature of American boyhood, real or imagined. One such selling method was to describe all the worldly advancement boys could gain through a career in science. A Gilbert manual from 1936 cited chemistry as "the [science] which offers the greatest opportunities of advancement, research and fame for those today who are interested in the fuller things of life."[58] Another tactic was to point to the in-the-moment enjoyment that boys might feel through their new hobby—enjoyment that might rival other, less interactive "modern" entertainments, such as going to the movies. In 1923, a Chemcraft manual cited an unattributed quote from a child who said, "I would rather do experiments with my CHEMCRAFT set than go to the show."[59] Companies, realizing that their customers, growing up in the age of moving pictures, radio, and pulp magazines, had other leisure opportunities, sought to compete accordingly.

Companies tried to sell children on the emotions surrounding discovery, in particular emphasizing what were always described as the "thrills" of scientific activity. The Microset manual invited the reader to continue to ask questions, promising that "you will enjoy the same thrill in finding the answers, that all scientists experience—the thrill of discovery and accomplishment that keeps them studying and exploring throughout their lives."[60] A 1940 Gilbert catalog had an introductory note from its founder, who wrote: "To our forefathers, adventure meant blazing a trail through the wilderness, fighting off Indians and wild animals, and conquering the forces of nature. . . . Today the world's most thrilling adventures lie in the field of science—engineering, chemistry, electricity, microscopy, and other forms of scientific activity." These adventures, A. C. wrote, could compete with the storied exploits of days of yore: "There is as much romance in discovering the secrets of strange chemicals as the secrets of strange lands . . . there is as much glory in conquering the wild forces of electricity as in conquering wild tribes."[61] These comparisons, which so strongly echoed the racist and imperialist tropes of adventure stories popular during A. C. Gilbert's own Progressive Era youth, moved action indoors, privileging feats of the brain over feats of brawn. The argument also stressed the easy availability of these thrills and romances, even for boys who lived in urban or suburban places and dreamed of a nineteenth-century pioneer life. Boys using chemistry sets could prove that they were still conquerors, even if modern life was not conducive to the blazing of physical trails. Finally, in invoking "our forefathers" (the pioneers and fighters of Indians), A. C. hailed white American boys as his audience.

A final appeal was the appeal to patriotism. "Chemistry is more closely interwoven with the industries of the world than any other science, and the country which leads in chemical industries will ultimately be the richest and most powerful," one undated Chemcraft manual explained. "It will have the fewer [sic] waste materials, it will have the best manufactured articles, its food will be the most nourishing and the cheapest, and it will possess the secrets of the most powerful explosives, the hardest steels and the mightiest engines."[62] This appeal to patriotism would become far more common during World War II and the postwar era, when, for example, participants in the first Westinghouse Science Talent Search wrote essays on the importance of science in defending the United States from attack, or "Rocket Boy" Homer Hickam reported feeling like he and other high school students were "being launched in reply" to Sputnik.[63] Viewing the rhetoric about the scientific vocation and patriotism that was present in these chemistry sets, we can see the foreshadowing of these postwar appeals to young people's sense of duty.

THE HOME CHEMISTRY LAB AND THE BOY'S SOCIAL WORLD

Like Thomas "Al" Edison in his railroad car, boy experimenters needed to have a separate place in which to carry out their experiments. Manufacturers of chemistry sets imagined that boys would be able to secure such a space somewhere in their parents' homes or right outside of them. Manuals offered recommended procedures for setting up home labs, suggesting locations and raising factors to consider when picking a spot; electrical connections, running water, and good lighting, they counseled, would all improve the experience. The more complete manuals also offered tips on constructing lab benches, shelving, and lamp shades that would modify lighting to create optimal lab conditions.[64] Chemcraft's official magazine published letters from Chemcraft clubs with evidence of the actual locations of labs. "Chief Scientists"—leaders of their respective clubs—wrote in that they had "a clubhouse laboratory" (James Rodgers, Grove City, Pennsylvania); had been given "a fifteen square foot space in our basement" (Louis Schlitz, St. Paul, Minnesota);[65] had "partly partitioned off the basement with [a] bulletin board, [a] case for apparatus, a regular desk and a table large enough to accommodate three or four members" (Alois Dettlaff, Cudahy, Wisconsin); were "making plans for a larger and better equipped lab on the roof of our house" (William Hudson, New York, New York);[66] and had "received permission from our country [*sic*] commissioners to use the voting house in our district as our club headquarters" (Joe R. Mahr Jr., McKees Rocks, Pennsylvania).[67]

How much did parents know or worry about the experiments going on in these basements and clubhouses? After all, not all work that happened in these spaces was motivated by the "right" kind of curiosity. Manuals and official publicity rarely referred to the more destructive results of some experiments, but boys who grew up during this time have fond memories of a certain degree of mischief. As engineer James Waters (b. 1925) told an interviewer about his own childhood chemistry set, "I did a few experiments, but I was more interested in making something that went 'poof'!"[68] John Lienhard describes a laundry room experiment in the 1930s in which he created a musket ball by pouring molten lead into a mold; during one series of experiments, he cleaned the mold with water, then returned to pouring in lead without bothering to dry the mold out thoroughly. "Molten lead blew outward in all directions, spattering my face and narrowly missing my eyes," he writes. "I was hurt and frightened, of course, but ongoing injury was part of the game."[69] Companies took a few missteps in this respect: in 1950, Gilbert admitted to an interviewer that the earlier Gilbert sets had contained what the article called "a certain chemical" that, in Gilbert's words, "when placed in a test tube with certain other compounds, it would give off a tiny, almost imperceptible flash." Gilbert went on, grumpily: "The only way a boy could possibly be hurt by it was to hold the tube against his eye ... which is just what two or three of them did do." Despite Dr. Johnson's entreaties that the chemical was necessary for learning, the company stopped including it in sets — an early amendment of equipment in chemistry sets that foreshadowed later self-modifications made by the industry during the 1970s and 1980s.[70]

In manuals included in sets during the interwar period, companies tried to persuade children to be careful by telling them that carefulness was scientific and adult. Precision and forethought, the manuals argued, were the hallmarks of a "real" experimenter, who would likely not experience explosions in his lab. In its "introduction," a 1940s Chemcraft manual tried to disabuse children of the idea that explosions were inevitable, letting them in on what the language tried to cast as a little-known truth: "Chemistry is sometimes looked upon as a dangerous profession, but this is not the case. Contrary to an

The cover of the Porter Chemcraft Chemical Outfit, circa 1928.
Courtesy of the Chemical Heritage Foundation.

old popular idea, a chemical reaction does not necessarily result in an explosion."[71] A child who was hurt or poisoned by a chemistry set, the logic went, was not following scientific procedure. An undated Chemcraft manual juxtaposed the foolish child with the careful scientist: "Chemistry is a science of systematic procedure and control. Everything is done for a definite reason. Do not combine any other reagents, substances or materials with the chemicals contained in this Chemcraft Outfit. Do not find fault with Chemcraft because of incorrect results that will follow chemical combinations other than those specifically stated in this Book." This manual flattered the rational reader with a warning: "Chemcraft is not recommended for children who are unable to read and understand this statement."[72] In making this distinction, Chemcraft provided a chance for the child to prove himself as "old enough" — as capable of handling responsibility and acting rationally.

The companies' approach to mitigating danger did not include adult supervision. Collaboration between parents and children in experimentation was not an expected factor in kids' home science endeavors during the 1920s, 1930s, and 1940s; far more common were cross-peer initiatives, exemplified by Porter's broad network of Chemcraft Science Clubs. Science clubs were a way for the toy companies to encourage the peer activity that the set boxes depicted; these organizations also offered companies a way to keep track of their patrons and encourage their investment in a corporate brand. While extracurricular science clubs run by nonprofit entities such as New York's American Institute and the American Association for the Advancement of Science proliferated in the late 1920s and early 1930s, they followed the commercial science-toy companies into the field.[73] The front of an undated Chemcraft set is a vision of the group model of experimentation. Here, a group of boys in coats and ties works with Porter-labeled chemicals in what seems to be an impromptu clubhouse. The presence of a benevolent moon outside the window indicates that the boys are here on their leisure time, which, coupled with the chiaroscuro shading of the boys' faces, reads as magical and mysterious, echoing some of the instruction manuals' focus on the alchemical past of this science. Two boys do the experimenting, while two others read booklets (possibly instruction manuals). A diploma on the wall may indicate membership in the Chemcraft Science Club or may simply be meant to point to the boys' interest in external markers of achievement. This was an idealized vision of science, as well as of youth; here, camaraderie and individual effort mingled to create a portrait of dedicated pure inquiry.

Companies moved to foster community among customers, promoting clubs almost as soon as they started selling chemistry sets. Porter started its Chemcraft Science Club in 1918. Clubs received the *Chemcraft Science Magazine*, which featured club news, articles about famous scientists, question-and-answer columns about chemistry, and order blanks for all the Porter supplies. Club branches, which took individual names and named "Chief Scientists" as leaders, wrote in to *Science Magazine* to report on the progress of experiments. In 1940, for example, Richard R. Bailey, of Ocala, Florida, reported on the doings of his group, which had taken the name "The Pasteur Science Club." The Pasteurians were "experimenting on spiders and Florida moss" and looked forward to big events in the future: "We are planning very soon to dissect a rat while under the influence of ether. During this dissection we intend taking out one kidney and fix the incision to the best of our knowledge. If he lives we will take a record of his actions."[74] Like child members of other national "clubs" sponsored by corporations, such as fans of Little Orphan Annie or Tom Mix, Chemcraft Science Club members could identify each other through indicators of membership. In 1937, members of the clubs were sent badges, "to be worn in plain view at all times," "to show that you are interested in the progress of the world through the development of science." Senior members received a password, "written in invisible ink on your membership card."[75]

Porter advertised the science clubs heavily and used the information it received from members in product planning and advertisements. In 1917, the Porter Chemical Company placed an ad in *Playthings* touting the sales record of the past year — "95 per cent of the stores SOLD ENTIRELY OUT" — and promising that "stock carried over will be moved early this year by the 20,000 members of the Chemcraft Chemist Club; boys and girls who are pushing CHEMCRAFT all the time."[76] Chief chemist Harry K. Phillips, of the Pine Tree Chemcraft Chemist Club, appeared in an advertisement in *Playthings* in 1922, admonishing the buyers reading the publication: "If you want to sell something nine out of ten boys and girls are interested in, sell CHEMCRAFT. It's the only GOOD chemical set there is; gives the biggest value for the money, and it's real chemistry!"[77] Phillips appeared in his laboratory, striking a casual pose, and looking self-assured and happy.

Beyond their advertising advantages, clubs run by chemistry set manufacturers clearly served as platforms for selling merchandise to participants. In 1940, a Gilbert catalog suggested that a Boys' Engineering and Science Club needed as many different types of Gilbert toys as possible, framing the club as a way for the younger generation to pool its access to parental largesse. "Here's how to go about it," the catalog counseled. "You and the other boys look over all the toys described in this book. Then one boy asks his Dad for an Erector set for Christmas . . . another boy puts a Gilbert Chemistry set on his Christmas list and so on until you have a complete assortment of everything you want. . . . Why not talk this over with the fellows today and make your plans right now?"[78] While consumers buying from Gilbert and Chemcraft could purchase instructional manuals separately from sets and (hypothetically) could build their own collections of chemicals through individual purchase (as home chemistry experimenters had been advised to do in the nineteenth and early twentieth centuries), the companies clearly preferred to have children buy their sets.[79]

Companies selling chemistry sets often played upon boys' emotions of envy and the hierarchies of their peer groups in pitching their products. Advertisers who wanted to market products to boys during this time often spoke among themselves of trying to win over the "gang leader," considered to be the most charismatic of the boys in a particular neighborhood. Legitimizing children's feelings of consumer envy, such companies appropriated these feel-

"Harry K. Phillips Has a Message." From *Playthings*, March 1922.
Courtesy of The Strong®, Rochester, New York.

ings as a selling point in advertising.[80] Peer approbation, and a place as "gang leader," was a big part of the "thrills" that Gilbert science sets promised their young buyers. In an advertisement for his early 1920s science sets, Gilbert wrote, "The boy who knows about different types of engineering—electrical, chemical, structural, etc.—the kinds that are covered by Gilbert Toys, is the type of boy who will be a leader among his fellow boy friends. He is the boy whom the rest of the boys look up to, and they only do it because they appreciate that he has a knowledge of different things which they don't understand."[81] Sometimes ownership of a set would be enough to incite this type of jealousy—no need to possess difficult scientific knowledge. Advertising Gilbert's telephone and electrical sets in 1940, catalog copy touted the sets as "among the finest achievements of the Gilbert Hall of Science—sets that you will be proud to own and that will make you the envy of every boy in your crowd."[82]

Companies pitched chemistry clubs to boys as a way to learn group dynamics and managerial strategies, as well as science. An early Chemcraft Chemist Club bulletin made the argument that "a well-organized chemistry club, with an interesting program of activities, will be of great value to the members not only in increasing and broadening their knowledge of chemistry, but also in teaching the fundamentals of group activity and parliamentary law."[83] Chemcraft suggested that the bylaws of clubs provide for a point system, which would allow members to gain points for such activities as "making a trip to some industrial plant or laboratory and reporting to the club" (10 points), "knowing the names of ten great chemists" (10 points), or "making a great chemical discovery (as recognized by the club)" (25 to 50 points).[84] William Rogan, chief scientist of the Beaver Science Club of Bangor, Maine, wrote to the Chemcraft Science Magazine in 1939 that his club had instituted a rank system. Members started as "pledges," moved to "second class," "first class," and then "veteran experimenter." Rogan added: "When a person has achieved or reached the highest possible rank, he or she is given a party or banquet in his or her honor with 'after dinner' speakers on subjects of science"—an event reminiscent of similar functions held by adult social clubs such as the Rotarians or the Elks Club during this period.[85] The magazine sent to members of the Chemcraft Science Clubs offered managerial advice to the chief scientist. "Why not use January for the re-organization of your club and material?" a typical editorial asked. "Appoint a committee to work with you in laying out a program of study and activity for the month."[86]

Some clubs even made money with their chemical activities. The Yonkers Science Club, headed by chief scientist Donald Rhyns, manufactured ink and sold it, "at five cents a bottle. The money collected is being used to purchase new equipment."[87] A common money-making endeavor was water testing; James Rodgers wrote to the *Chemcraft Science Magazine* in 1937 that his club was making enough money from testing in its members' neighborhood to buy "a complete volume on chemistry and chemical training."[88] The fact that these Chemcraft clubs channeled the money from these endeavors back into science meant that their entrepreneurial spirit could be seen as constructive, forward thinking, and rational, rather than acquisitive or greedy. Gilbert also provided an annual monetary prize for chemistry achievement; in a 1935 *Boys' Life* ad, Stanley Stewart, of Waycross, Georgia, was depicted in the midst of demonstrating his "Green Fairy Fire" experiment, with three friends looking on. The ad informed the boy reader that Stanley had received a $100 prize from Gilbert, adding that other "Gilbert boy chemists" had won prizes of money from other unnamed sources: "One boy received $150 for perfecting a candle that burns with a blue flame. Another developed his own formula for making soap, sells it and earns big money."[89] While later models of financial compensation for children's scientific work, such as the scholarships given by the Westinghouse Science Talent Search, were tied to the child's continuation of scientific learning, the Gilbert prizes were external to the formal education system; instead, they were indications of the young chemist's future earning power and immediate sources of excitement.

One more money-making endeavor that the young chemist could undertake, with or without his club, was the "chemical magic" show. The scientific magic show would have been familiar to the young owners of chemistry sets in the interwar years; entertainments, including popular scientific demonstrations, the electric "wonder show," the vaudevillian mesmerist, and the magician, mingled science, pseudoscience, and showmanship in various measures in nineteenth- and early twentieth-century popular culture.[90] Beginning in the 1850s, instructional manuals for magicians included lessons in "chemical magic," and British and American popular literature about chemistry taught readers how to perform "chemical tricks" for (as one 1894 pamphlet put it) "home amusement and instruction."[91] The chemical-magic manuals that owners of interwar chemistry sets could send away for were part of a long tradition of textual instruction in chemical spectacle.

More than any of the other activities suggested by interwar chemistry set manufacturers, the promotion of "chemical magic," an activity that put boys at the center of a "show," demonstrates how strongly the makers of chemistry sets meant to ally their users with a self-consciously rational and modern set of beliefs and commitments.[92] As American stage magic in the time of Houdini aligned itself emphatically with the rational against the pretensions of spiritualism, chemistry sets gave boys the means to harness the "scientific" in service of the spectacle, casting the boy experimenter as the knowing individual at the center of a web of illusions meant to entertain and mystify his friends. Most toy company brochures promoting chemical magic took a historical view of the practice, describing the origins of chemistry in alchemy and emphasizing the idea that the modern boy in the know could use the superstitions of the past for profit. In a Chemcraft *Chemical Magic* manual of 1937, the capsule history of alchemy at the beginning of the book noted: "In those days people were very superstitious, and so the alchemists who had learned to bring about such wonderful [chemical] changes in materials came to be regarded as wizards or magicians." But, the manual continued, the modern reader should give the people of the past the benefit of the doubt: "Numerous achievements of present-day scientists seem like miracles of magic until we understand the scientific principles on which they depend."[93] However understandable the enjoyment of mystification might be, the manuals made clear, boys performing as "chemical magicians" should know that these tricks did not constitute actual science. Chemcraft's magic sets drew a clear line between the tricks contained within and "true" chemistry: "The things which can be done with the contents of this outfit are not strictly chemical experiments, as are those done with CHEMCRAFT sets, but are especially developed 'stunts' or tricks of chemical magic having a puzzling and sometimes startling effect."[94] The boy should know the difference.

Many tricks suggested in chemical magic books involved colors: changing water to different colors; coating paper with chemicals and then writing on it with water; creating different kinds of "sympathetic ink" ("such inks have frequently been used by spies in time of war for conveying military secrets through enemy territory").[95] Manuals also offered instructions in the transformation of household objects, so that a child could create "elastic" eggs or chicken bones through soaking them in vinegar or make "disappointing matches" painted in sodium silicate that would strike and then almost immediately go out.[96] "Diabolical odors" was a promising subsection in one manual, offering instructions for generating a "disagreeable odor" (sulfur), "a magic odor that will revive fainting persons," or the apparently diabolical "odor of violets."

These chemical magic shows, suggested and scripted by the manuals produced by toy companies, served as training grounds for boys who wanted to learn how to command a crowd. A. C. Gilbert credited his youthful magic hobby with teaching him habits of perfectionism and persistence; he often pointed to a boyhood incident, in which he had attended a magic show and been allowed on stage to perform some of his own illusions, as the genesis of his own seemingly boundless self-confidence.[97] Toy companies took the cultural phenomenon of the magic show and recast it with young people as showmen rather than spectators, offering the boy customer a chance to be the fooler, instead of the fooled. In their performances, the boy chemists would show their friends and family how great the differential between their knowledge and the knowledge of those who remained outside the laboratory had grown.

Stagecraft was an important part of the instruction offered by these manuals, which gave readers lessons in controlling an audience. The Chemcraft manual of 1937 suggested that the reader "make-up as an Alchemist" to "add very much to the interest and impressiveness of your entertainment," not least because "it will make the show appear more professional and help the performers to keep from laughing while enacting their mystic roles."[98] This coaching on presentation was also about commanding a stage presence. "Always speak SLOWLY and IMPRESSIVELY with DRAMATIC pauses at the proper places," Chemcraft advised. "Ignore any questions, remarks, or other attempts to interrupt you."[99] At the beginning of the show, the "alchemist" was counseled to "speak a word of caution": "Know ye that the spirits of the Alchemists of Old, by whose aid I these wonders do perform, are proud spirits and permit no ridicule or unbelief. If therefore, be there any amongst you that is an unbeliever in the mysteries of science, I beg that he now withdraw, lest ill results betide him here."[100] This injunction was meant to set up the presenter so that any future interference from the audience could be dealt with sternly, perhaps by the use of a sound effect to indicate the displeasure of the spirits.

Showmanship, as an integral part of science, reinforced divisions between the "head alchemist" and those who assisted with and watched his magical demonstrations. Manuals addressed issues of management, offering ideas for ways that boys could direct the friends who would help put on magic shows. Chemcraft suggested that the head "alchemist" could "appoint one individual to have complete charge of publicity."[101] Despite this delegation, the boy in charge must ride herd on his compatriots: "Everyone connected with the active production and management of the show should sell his share of the tickets. Have everyone go after this matter and encourage them to sell all that they possibly can."[102] The single reference to a nonwhite person in all the manuals that I examined clarifies these divisions even further, making it evident that the white boy at the center of the stage could use historical power relationships to evoke a further illusion of his mastery. The 1937 Chemcraft manual, reprinted several times in later years,[103] suggests that the magician's assistant be costumed as "an Ethiopian slave," with "face and arms blackened with burned cork which will wash off easily when the performance is concluded." When the alchemist was to address the slave, he should refer to him as "slave." However, "if you prefer, the blackening of the face and arms can be omitted and the Assistant can be called 'Apprentice' instead of slave. By all means assign him a fantastic name such as Allah, Kola, Rota or any other foreign-sounding word." Clarifying the normative expectations of the author, despite the presence of this alternative suggestion, the rest of the instructions refer to this assistant as "slave."[104] By stripping the relationship between the "alchemist" and the "slave" of any recognizable historical context, the "chemical magic" instruction booklets skirted actually existing American race relations while reinforcing once again the expectation that the boys in charge would be white.

By providing these suggestions for effective stagecraft, the manuals instructed boys in the control of an audience, placing science within a larger lesson about performance and authority. Both Gilbert and Chemcraft saw chemistry sets as rational entertainments that fit into their vision of boyhood as a time that young men would spend cementing social bonds, learning to direct "active-mindedness" into productive curiosity, and acquiring habits of entrepreneurship and management. In recommending the construction of labs, the founding of clubs, and the mounting of shows, Gilbert and Chemcraft imagined science as an integral part of the boy's social world.

GIRLS NOT ALLOWED

Everything about the way the chemistry set was marketed and discussed signaled that female members of the family—sisters—should not assume the identity of "young scientist." Sometimes this was an explicit exclusion: the A. C. Gilbert Company's motto was "Hello Boys!" Other times, the divide was subtler. This cultural work echoed larger trends in the adult world. The professionalization of science during the late nineteenth and early twentieth centuries had the net effect of excluding women from the scientific professions; women trained in science, even those who had received Ph.D.'s, often ended up working as laboratory assistants, "computers" crunching data for projects run by men, or short-term employees. In chemistry, the demands on the industry during World War I, when American companies had to take over the work formerly done by German industry, were so great that some women had a chance to enter industrial chemistry for the first time; however, in a foreshadowing of what happened to female defense workers after World War II, when the war was over most of these women resigned their positions.[105] In the post–World War I era, science education in schools moved away from nature study and toward the physical sciences, a move accompanied by the exclusion of girls from "serious" science.[106]

Toy industry publications wrestled with the question of gender roles, not wanting to tread on current mores, while also hoping to reach as wide an audience of customers as possible. In 1921, *Playthings* ran an article titled, "Proposing, a New Field for Toys — Sell Toys to the Girl!" To justify what might seem to the reader a purely commercial interest in this expansion, the article went on, "All of us realize that girls and women of today, for better or worse, are moving into masculine fields, doing what has heretofore been men's work. But all of us do not realize that thousands, no, millions, of little girls are invading the play field of the little boys." The author posited that girls allowed to use boys' toys might be "trained for the business world so that they will not be helpless if there is not to be a domestic life" and mused that "even if the little girl never directly knows the business world, she will be a far better helpmate to her husband if she has been broadened by real boys' toys."[107] It would be better for the company to accomplish this "broadening" by chance than by design, however; *Advertising and Selling* reported in 1946 about A. C. Gilbert's approach to advertising: "Gilbert's advertising is addressed solely to boys, not to parents or to boys and girls. Girls don't mind this; boys would shy away from anything advertised for boys and girls as sissy stuff."[108] Companies making chemistry sets sometimes produced kits made specifically for girls; an example was Porter's attempt to reach girls during the twenties, SachetCraft, a kit one could use to make perfumed pillows (this outcome, the company insisted, "delights the girls").[109] Reviewing the product, *Playthings* wrote in 1920 that "an article for girls with which they could do something practical and amuse themselves at the same time has long been a problem, and 'Sachet Craft' seems ready to supply the answer."[110] Gilbert, for its part, produced a nursing kit during World War I that contained "a complete uniform, illustrated primer, and first aid equipment."[111]

The approach that chemistry set manufacturers took to serving girls echoed the treatment of young women in industrial "fairy tales" and storybooks marketed to modern children. The central "finder-outer" in interwar industrial story books tended to be male; any female characters that appear in the narrative serve as auxiliaries to the action, their interest in the proceedings of the factory tour or the lecture circumscribed. In narratives whose subtext was always, "It's good to be curious about everything," girls are depicted as being interested mostly in things that are nice to look at or that smell good; their aesthetic reactions take the place of any considered analysis of the implications of the material that they see in front of them. Authors Maud and Miska Petersham began their story of the discovery of gold with a hypothetical little "cave girl" who sees something in a stream and decides she thinks it is pretty ("bright like the sun"), then asks her father to make it into a necklace.[112] Sister Mart comments on crushed slag in William Clayton and Helen Sloman Pryor's *The Steel Book* (1935): "It's pretty again after it's crushed. I like the way it sparkles."[113] Girls were easily distracted from inquiry. In the Pryors' *Streamline Train Book*, Nancy sits down to read a book, while her friend Ted stays by the window, finding the passing landscape more interesting.[114] These books, touted as appropriate for modern children, depict girls as diverted by aesthetic concerns or by such frivolous occupations as fiction or magazines, whereas male interest is most happily occupied by the workings of reality.

In the groups of children depicted on chemistry set boxes or in manuals, girls took the position of spectators, watching with interest and admiration as their brothers interacted with the components of the chemistry sets. Science textbooks published for high schoolers during the 1950s contained few images of women or girls, and the images that they did contain were often of homemakers or nurses.[115] Evidence from these sets and manuals shows that these informal educational tools enforced similar gender binaries in the 1920s, 1930s, and 1940s, visually depicting manufacturers' expectations about the social relationships that these sets would promote within child peer groups. Just as the girls in interwar "information books" were interested in science insofar as science could produce pretty things, many of the images of home chemistry labs from the same time period that do include girls cast the boys in the role of "showman" or "actor," while the girls are approving and pleased onlookers. The back cover of a Lionel-Porter manual first published in 1937 depicted a lab scene that was typical of one mode of approach to picturing girls: the token girl, allowed seemingly because of her singularity. In a 1928 advertisement for Chemcraft sets, included inside a Chemcraft manual, illustrations depicting the bigger, more expensive sets feature high-school-age boys with girls, in which the boys appear to demonstrate or gesture toward the sets, as though showing off their new toys.[116]

Home chemistry labs — sometimes in truth, and often in legend — existed in opposition to mothers' efforts at imposing domestic order, a fact that amplified the perceived maleness of the space, further excluding girls. Changes in the spatial arrangements of the middle-class American family home made dedicated masculine spaces for science possible. "Children's rooms" became a common feature of American family homes only in the period from 1890 to 1930.[117] By the 1920s and 1930s, it was the fashion to decorate boys' rooms with military motifs or to outfit these spaces with accoutrements intended to develop future careers. (Girls' rooms were not themed in the same way.)[118] Other boys, like the many Chemcraft scientists quoted above, were given extra bedroom spaces to practice their hobbies.[119] The home space known as the "laboratory," as well as the scientific "research" that a boy carried out in this space, was a part of the house that was both physical and symbolic. The lab and the absorption in science that it signified enabled a young man to establish a certain distance from the rest of his family: older and more responsible than younger siblings, while at the same time more forward thinking and informed than his parents. This kind of distance was the cultural province of middle-class, school-age boys, who had long expected to be allowed to romp outside with groups of male friends, while girls were kept indoors.

The boy's lab was also a space where domestic disorder could thrive — another refuge of the appealingly wild "bad boy," a figure in American popular culture since after the Civil War.[120] In fact, that disorder is one of the most treasured aspects of the collective cultural memory of childhood chemical experimentation. The little tale of youthful destruction is a staple of the twentieth-century male scientific autobiography. Physicist Richard Feynman recalls throwing a flaming wastebasket out of the window so that his mother would not find out that he had been playing around with a spark coil; neurologist Oliver Sacks recounts an incident in which he and a friend so thoroughly polluted a friend's basement with chunks of fermented cuttlefish (products of an attempt at preservation, gone badly awry) that the house was rendered uninhabitable for weeks.[121] In such stories, the gender dynamic — frustrated mother, who has to clean up this mess; incorrigible, yet brilliant, son — repeats itself again and again. The message is clear: scientific advancement relies upon boyish mischief. The wholesome transgressions that took place in the lab would have been much less socially legible if committed by a girl.

Paradoxically, unsupervised home science labs for boys took hold just as safety expectations for the rest of the home were on the upswing. In the first half of the twentieth century, safety movements, which were mostly aimed at improving the health of workers and pedestrians in mechanized spaces of factories and city streets, made some inroads on questions of home safety. The advocacy group National Safety Council (NSC) began the first of many public awareness campaigns in 1912, addressing industrial safety first, then public safety, in trying to reduce numbers of auto accidents. The NSC launched its first household safety campaign in the 1930s.[122] Also during the 1930s, the Red Cross mounted a major campaign to heighten awareness of home safety, distributing over 26 million home inspection forms to children, who would bring them home to their parents.[123]

These efforts emphasized the need for responsible action on the part of the consumer. In particular, during this period, advice on household safety increasingly emphasized the mother's duty in keeping children, husbands, and visitors safe.[124] The transition between the Victorian perception of the "home-as-haven" to "home-as-dangerous-place," part of a new Progressive Era awareness of environmental risk, led to calls for women to become "home safety managers." This role aligned with the period's emphasis on efficiency and scientific practices in home economics — a stripped-down, "modern" response to the fripperies and decorations of the Victorian home.[125] Women — who, it was understood, knew their houses best — should also know how to avoid dangers like slipping and falling on wet floors or fire hazards such as poor electrical wiring or piled-up clothing.[126] This expectation makes it all the more remarkable that scientific experimentation in the home in the interwar years was so emphatically unsupervised — and shows how strong the ideal of boyish science was.

In many stories of childhood experimentation held up as examples for children, boys challenged adult women's expectations of behavior and attention. An 1867 Marcus Stone painting of inventor James Watt as a young boy playing with a teakettle was a common fixture in the children's encyclopedias of the 1910s, 1920s, and 1930s.[127] Author Gertrude Hartman, writing about the portrait, describes the small experiment and then imagines the response of an "aunt" who "reprove[s] Watt from what she thought was trifling" behavior: "James Watt, I never saw such an idle boy. Why don't you take a book and employ yourself usefully? For the last hour you haven't spoken a word, but have just taken off the lid of that kettle and put it on again. Aren't you ashamed of wasting your time that way?"[128] A child reader would know that this aunt was the foolish one; this reversal is a pleasing testimony to the importance of instinctual curiosity as opposed to the kind of channeled and directed learning ("usefulness") that made sense to adult female onlookers. Expectations that girls would be "useful" resulted in the recommendation that they pursue domestic science rather than the less immediately helpful pursuits of chemistry or physics.

CONCLUSION

Chemistry's properties—its mysterious nature, its presence in the everyday object and the home, its potential to confer power upon those who might master it, and its shrinkability—rendered it uniquely suited for packaging for home use by children. During the interwar years, manufacturers of chemistry sets offered parents a vision of their children's modern futures while telling boys that chemistry could provide status, thrills, and feelings of mastery. Through these sets, the home became a location where science happened, and the boy stepped firmly into his twentieth-century role as the home's scientist-in-residence.

The boy's chemistry laboratory was particularly significant in its departure from previous models of in-home science education. During the eighteenth and nineteenth centuries, the parlor and the home library in middle- and upper-class homes were places where science—demonstrations, the display of cabinets full of specimens, discussions of books—happened; these interior, yet familial spaces were locations where younger and female family members might encounter science in a domestic setting, absorbing a fascination with natural history, astronomy, or experimentation from their older male relatives.[129] The shift from this model to that of the home chemistry labs described in this chapter is a transition that moves the location of cutting-edge home science from the older generation to the younger and from the familial to the individual. The exclusivity of the boys' "science clubs," which were meant as a site for independent investigations undisturbed by adult guidance, reflected a new cultural belief in the appropriateness of separate youth activities. Returning to Natalie Angier's observation about twenty-first-century American culture—"childhood is the one time of life when all members of an age cohort are expected to appreciate science"—this shift seems particularly significant.[130] The marketing and packaging of chemistry sets appealed to adult eyes, showing the joy of childhood investigation, and the idea of the chemistry set appealed to adults who liked to think of their children as "modern"; along the way, science was increasingly associated with boyhood.

The history of the selling of chemistry sets during the interwar years provides an interesting prequel to the Cold War history of worries over science commitments and a lack of "manpower." Adults who grew up with the interwar culture of chemistry sets, such as science fiction author Robert Heinlein, wondered how childhood enthusiasm for science had been lost, pointing nostalgically to their own enthusiastic boyhood participation in experimentation and contrasting this attitude unfavorably with the peer-oriented, popular-culture-obsessed youth of the present day.[131] The history of the interwar promotion of chemistry sets reveals there was more continuity between the prewar utopia of experimentation and the postwar regime than such worries would indicate. Even in the supposedly pure interwar years, manufacturers relied on peer networks, appeals from popular culture, and an acquisitive materialism to promote science practice.

Embryo Scientists

Finding and Saving Postwar "Science Talent"

In 1958, *Life* magazine ran a multi-issue series on the "crisis in education." The first article in the series was headlined, "Schoolboys Point Up a U.S. Weakness." The piece compared "likeable, considerate, and good-humored" Stephen Lapekas of Chicago with "hard-working, aggressive" Alexei Kutzov of Moscow, following each boy through a week of school and extracurricular activities. The article and the series are highly ambivalent artifacts, exemplifying the severe identity crisis that the postwar United States faced when it came to the topic of education. While acknowledging that the American system did have the advantage of producing flexibility and "qualities of leadership" in students and that the Russian system "develops rigidity and subservience to an undemocratic state," the editors of *Life* produced a pictorial narrative that showed Stephen, along with those around him, to be soft-minded, self-indulgent, and pleasure-seeking, whereas the more anxious and buttoned-up Alexei lived for intellectual accomplishment and hard-won approval from his teachers.[1]

If the public of 1958 wanted to know why the USSR had managed to reach space before the United States, *Life* found the answers in these teenagers' mental landscapes. In English class, *Life*'s photographers captured Stephen and a classmate reading aloud from a play, while a female student in the back row perused a barely hidden copy of the magazine *Modern Romances*. The photograph across the spread depicted Alexei, in a uniform coat, standing to be examined in a laboratory setting. While Alexei and his classmates are pictured arrayed in careful ranks around a science museum exhibit, paying serious attention to the words of the docent, Stephen and his laughing friends are only "momentarily diverted" by a biology project—prepared, as the caption noted, by another student at the high school who eventually submitted it to a citywide science fair, where it won first prize. In this school—and in American schools in general, it is implied—science-minded students stood outside the mainstream, creating objects only of momentary amusement to the more popular kids.[2]

Life complemented the visual representation of a happy and popular Stephen a few issues later with the article-length portrait of a "gifted" but "lonely" eleven-year-old from Iowa named Barry Wichmann. While describing Barry as a "wasted" talent, *Life*'s photographs illustrated how Barry's comparatively "richer" mental landscape and range of interests isolated him from his peers. A photograph of Barry reading the newspaper while his father watched television was accompanied by the caption: "Though fond of horror shows, Barry prefers reading to television, comments 'I think the habit of watching TV on Saturday night has a deteriorating effect.'" Barry reveals to the interviewers that he has often tried to cultivate particular interests, in order to fit in: "[Barry's] comments on baseball give a heart-rending insight into his need for acceptance. 'I've tried and tried to like baseball,' he says, 'but I simply can't like it.'"

In *Life*'s representation of Barry's days, science has a shadow presence. For Barry, science seems to lurk as a possible occupation that is just out of reach. A photograph of Barry peering wistfully around a door into the high school science lab is accompanied by a caption: "In past years he knew the science teacher well, came in lab often just to poke around at bottles and ask questions. But now there is a new science teacher and Barry is too shy to come in unless the room is empty."[3] If the contrast between Barry's attention to the paper and his father's attention to television shows how Barry's superior mind isolated him from popular culture and from the people around him, the forlorn peek around the door of the science lab illustrates how Barry's culture—one that does not offer science instruction to children so young and that fails to be flexible enough to accommodate special cases—isolates a "brilliant mind" from science.

According to one anxious vision of the mainstream culture of the postwar United States, the science-minded young person, yearning for serious fare, had become a tragic figure, betrayed by his elders.[4] Novelist Sloan Wilson's article in this *Life* series sketched what he saw as a typical scenario unfolding in high schools. "Many a brilliant youngster finds that his school has assumed the aspects of a carnival," Wilson wrote. "In one room pretty girls practice twirling batons. The sound of cheers is heard from the football field. The safe-driving class circles the block in new automobiles lent by an enterprising dealer. Upstairs funny Mr. Smith sits wearily on a stool in the chemistry lab trying to explain to a few boys that science can be fun, but who pays any attention to him?"[5]

Wilson invokes several specters of distraction: "pretty girls," sports, the consumer freedoms afforded by automobiles. His choices were typical of critiques mounted of postwar education, and his selection of the chemistry lab as the opposite space within the school—a place where a few adults mounted a rear-guard action for intellectualism, against the forces of peer culture—was also characteristic. For Wilson and for other American adults who grew up in the era of the chemistry set and the encyclopedia, the new prescriptions for well-adjusted teenage popularity were cause for panic. What would happen to the questing mind of the home chemist, radio boy, or aviation buff, adrift in a wasteland of football games and comic books?[6]

RESCUING BABY SCIENTISTS

The Westinghouse Science Talent Search (STS), conducted by the nonprofit Science Service, which put high school seniors through a series of tests and selected forty that it considered to be the best in the country, was established during World War II with the purpose of seeking out teenagers who could provide future scientific "manpower" to a nation that desperately needed it. In the late 1940s and the 1950s, as scientists and science promoters increasingly diagnosed a mismatch between postwar peer culture and STEM pursuits, STS winners were drafted to provide insight on the factors that had allowed them to stick to scientific pursuits in the midst of this perceived cultural change and to serve as a guiding light to other children wondering whether science could be for them. The tastes, hobbies, and habits of their younger days, when they were said to have gained their love of scientific activity, became the object of scrutiny. By analyzing and representing their psychological makeup and their life stories, adult onlookers tried to make a place for the love of science in postwar childhood.

The historiography of postwar science and education revolves around anxieties of recruitment, arguing that the Cold War, the advent of nuclear weaponry, and the "space race" meant that adults sought to interest children in science to ensure a supply of young scientists for future national supremacy.[7] What is often missed is that the dynamic of "recruitment" was one that incorporated a strong critique of the surrounding culture. Adults calling for recruitment sought to restore what they perceived as a lost cultural climate in which children who were interested in science — and managed to develop the independence, creativity, and single-mindedness perceived to be associated with scientific activity — would be rewarded.

While prewar promoters of science as kids' culture assumed that the match between science and American children was a natural one and celebrated the curious young museum visitor, reader, and basement experimenter, during the postwar era the archetypal young science obsessive suddenly seemed both hard to find and increasingly precious. Larger changes in the symbolic value of childhood may account for some of this shift. While early twentieth-century Americans looked to childhood as a symbol of the progress of the nation, relying on the language of evolutionary theory to envision children as representative of an ideal past and future, midcentury Americans used discussions of American childhood and youth to address contemporary psychological questions about the development of people within societies and especially the relationship of individuals to authority.[8]

In keeping with this shift, promotion of science play in the postwar era focused on the benefits of science in helping children maintain an individualistic point of view in the face of a conformist society. In the late 1940s and the 1950s, the Science Talent Search advanced a vision of serious-minded, idealistic, creative youth that stood in stark contrast to the juvenile delinquent or the peer-obsessed teenager. The STS was determined to show that its "finds" were not neurotic, isolated, or strange; in its representation of young scientists, it directly responded to the new pop-cultural figure of the alienated young intellectual.[9] Margaret Mead and Rhoda Metraux's 1957 study of a sample of high school students' negative attitudes toward scientists was often cited in the popular press as evidence for the disconnect between the government's goals in terms of science recruitment and the "normal" student's feelings about scientists.[10] Psychologists David C. Beardslee and Donald D. O'Dowd followed the Mead and Metraux study in 1961 with similar research conducted in a college setting, determining that undergraduates perceived scientists as "unsociable, introverted, and possessing few, if any, friends" and "believed to have a relatively unhappy home life and a wife who is not pretty." (In even more damning critiques, students said, "I wouldn't care to double-date with a scientist" and "Maybe it's not a good idea for him [the scientist] to be married.") As in the Mead and Metraux study, Beardslee and O'Dowd found in their research that the women surveyed did not name "scientist" as a favorable occupation for a future husband.[11] One underlying concern of these studies was with the stifling influence of peer opinion on the potential young scientist, who might not even reach the point of committing to a scientific path, because he did not want to be thought strange, unappealing, or undatable.

Activities such as the Science Talent Search were meant to encourage and nourish the fragile science-interested boy, who was perceived as under threat from anti-intellectual school environments, uncomprehending peers, and the prevailing conformist drift of postwar culture. In looking for examples of students who had survived this climate of what Science Service director Watson Davis called "complacent ignorance and negativeness"[12] and come out on the other end with a commitment to a life of scientific activity, the scientists and journalists in charge of the Science Talent Search in the late 1940s and the 1950s were looking for—and also constructing—a model of youthful balance that could resolve several tensions around public perceptions of science in the postwar era. Just as the young readers and attic experimenters of the interwar period brought modern methods of inquiry into the home, reinforcing parental comfort with the advance of industry, the intelligent and socially competent high school seniors found through the Science Talent Search would resolve cultural conflicts: between a model of education that demanded rigor and one that asked for "life adjustment"; between the precious commodity of individual interest and the "manpower" demands of the state; and between the intellectual and the social dimensions of science.

The young scientist, as represented by the STS, led a life that was located somewhat outside of the mainstream of peer culture. To adults looking for reasons to be optimistic about teenagers, the STS winner was admirable for that very reason. In the postwar era, the much-discussed rebellious adolescent (the juvenile delinquent, the Beat) represented American autonomy, a safe compromise between sociopathic independence and a stultifying conformity.[13] "Embryo scientists" (as the STS finalists were sometimes called), by virtue of their investment in what was perceived as an unpopular pursuit among their peers, were also rebels—of a productive type.[14] In 1950 sociologist David Riesman identified the "inner-directed" as a vanishing breed in *The Lonely Crowd*; his description of the qualities inherent to the inner-directed—a single-minded work ethic, an adherence to goals originating from within, a stubborn persistence and sense of responsibility—had much in common with the Science Talent Search's desiderata. In his 1956 study of students "lost to science," carried out for the National Science Foundation, Columbia's assistant dean Charles Cole made this connection explicit: "[The scientist] is perhaps the best example of Riesman's 'inner-directed' personality ... either because of his make-up or as a result of his work, the scientist tends to be reflective and self-reliant."[15] A young scientist would certainly have to have a strong gyroscope (in Riesman's terms) to have pursued such a vigorous, diverse, and independently planned scientific life as the STS demanded of its finalists.

At the same time, however, the STS wanted its prize students to be sociable and cooperative and to exhibit qualities of "leadership." Although the supporters of the STS would certainly not align themselves with the so-called life adjustment curriculum that was the focus of much outrage in the postwar era, they regularly adopted pieces of its rhetoric, calling for STS finalists to be "well rounded" and integrated into their communities. By representing the STS finalists as containing the best of the intellectual and social worlds, Science Service and other supporters of the STS tried to paint strenuous scientific effort as a normal, admirable, and sustainable aspect of an American postwar childhood. The STS criteria of independent effort, self-driven inquiry, ceaseless curiosity, and easy sociability was a complex model derived on contested ground, entering into postwar public debate over the meaning of science, education, childhood, and human nature. The STS project was a salvage effort, a public relations project, and a science experiment. The Science Talent Search cast itself as an exemplary site of science recruitment and as generator of prestige—one that sought to create a new peer culture based around science while reassuring the public of the simultaneous genius and normalcy of scientifically talented youth.

BIG SCIENCE, YOUNG AMBITIONS

American science in the postwar era underwent several significant structural transitions. After the success of the Manhattan Project and other wartime efforts, scientists found themselves simultaneously in demand and under suspicion. Such events as the 1947 establishment of the United States Atomic Energy Commission, the 1950 founding of the National Science Foundation, and the 1951 establishment of the President's Science Advisory Committee brought scientists in close contact with government decision makers and gave science a welcome measure of political clout and governmental financial support.[16] The decade following World War II saw the Department of Defense move into position as the biggest underwriter of American science; while public monies expended for defense research and development were fifty times greater during World War II than before, these levels were again matched by the end of the Korean War and then climbed even higher after the flight of Sputnik in 1957.[17] Some scientists worried that this level of government support would result in, as Harlow Shapley (astronomer, Harvard professor, and sometime judge for the STS) termed it, "domination by the military"—or, put more bluntly by mathematician Norbert Wiener, scientists were in danger of becoming "the milk cows of power."[18] In the postwar political climate, many left-leaning scientists, believing in the key scientific values of intellectual freedom and international cooperation, found themselves in conflict with the government's agenda.[19] Other scientists bemoaned the effects of the new level of financial support on research, believing that the ensuing bureaucracy burdened science, imposed too many obligations, and rendered it inflexible.[20]

As for public perceptions of science, in an age when science had seemingly accrued much prestige, scientists and science promoters were uneasy with several aspects of science's place in postwar culture. These included the representation of consumer technologies, such as cars, refrigerators, and fabrics, as the entirety of "science"; negative perceptions of the figure of "the scientist"; and a troubling development that could be seen as related to both of these trends: an increased unwillingness to financially support basic research.

Scientists and allies worried that the representation of "science" in popular media reduced science to its fruits without discussing the fundamental questions leading to these advances. The record of scientists' critiques of the 1939 World's Fair shows that these worries began before the war. Left-wing scientists in the immediate prewar period criticized this public celebration of "science" as overly focused on industrial marvels and handy gadgets and replacing serious discussion of science's social utility or the scientific method.[21] In the postwar period, advances in medicine and psychology were increasingly boiled down in the popular press to bite-sized facts and easily applicable, uncomplicated pieces of "advice for everyday life."[22] If "science" was increasingly visible in the popular media through simulation and synecdoche, the "scientist" fared little better, emerging as an ambiguous figure portrayed as alienated and incomprehensible. Individual scientists were increasingly visible; the architect of the Manhattan Project, J. Robert Oppenheimer, was a household name, and exponents of popular scientific efforts, such as rocket scientist Wernher von Braun, showed up in prominent media outlets such as

Disney's weekly television show.[23] Because of the success of large-scale war-time projects, as well as such medical breakthroughs as penicillin and the polio vaccine, scientists' work was nominally more prestigious than ever.

At the same time, in 1957–58, a Rockefeller Foundation–funded survey found that 40 percent of the American public thought scientists to be "odd and peculiar people."[24] The movie theaters were full of films featuring atomic monsters, human mutations, aliens uncovered by ill-considered forays into space, and, most of all, "mad" scientists. In coverage of science in popular magazines, criticism of science increased as research grew in scale and was more commonly publicly funded. Fears about espionage and memories of Nazi doctors' programs of human experimentation fueled calls for regulation as science lost some of its moral authority.[25] Engineer and historian John Lienhard wrote in his memoirs of those years: "If we had been celebrants of genius in the early twentieth century, genius now seemed poised to turn upon us. . . . The bloom of Modern was off the rose."[26]

Pointing to the public focus on the fruits of science, coupled with these uneasy feelings toward the figure of the intellectual/scientist, scientists and other onlookers diagnosed a troubling public incomprehension of the importance of pure (basic) research. In 1963, in his *Anti-Intellectualism in American Life*, historian Richard Hofstadter compared the American love for Thomas Edison to the public obscurity of nineteenth-century physicist, mathematician, and chemist Josiah Willard Gibbs, celebrated in Europe but unable to find any public fame in the United States. Hofstadter used Gibbs's example to prove his argument that Americans prized intelligence (which he defined as inventiveness and pragmatism) while ignoring intellect (theoretical acumen).[27] Many scientists felt pressured by the new sources of research money to produce certain results; some made personal and impassioned arguments for scientific freedom.[28] In his 1945 report to the president on a "program for postwar scientific research," *Science: The Endless Frontier*, engineer and presidential science adviser Vannevar Bush stressed the need to fund research "performed without thought of practical ends," arguing, "Basic research leads to new knowledge. It provides scientific capital. It creates the fund from which the practical applications of knowledge must be drawn." This was an issue of national security, Bush argued, since "a nation which depends upon others for its new basic scientific knowledge will be slow in its industrial progress and weak in its competitive position in world trade, regardless of its mechanical skill."[29] Despite this practical argument in support of funding for basic research, we have several pieces of evidence bespeaking a 1950s climate hos-

tile toward these kinds of investigation, including the famous observation by
Eisenhower's secretary of defense Charles E. Wilson (who cut the Defense
Department's funding for research and development): "Basic research is when
you don't know what you are doing."[30]

Worries about scientific "manpower," an aspect of the postwar scientific
landscape that directly affected young prospective scientists' lives, can be
identified as early as the 1910s and 1920s but expanded greatly in the immediate postwar era.[31] In the *Science: The Endless Frontier* report, the Committee on
Discovery and Development of Scientific Talent (of which Science Service's
Watson Davis was a member) condemned the Selective Service's policy of
recruiting most graduate and undergraduate students of science into the services, without consideration as to the future.[32] The committee reported the
concerns of scientists inside and outside the academy. Dr. Charles L. Parsons,
of the American Chemical Society, argued that "today, we are drying up pros-
perity at its source. . . . Public opinion of the future will view with amazement
the waste of scientists in World War II." Dr. Charles Allen Thomas, director of
the Monsanto Chemical Company's research laboratories, believed "scientific
suicide faces America unless immediate and adequate steps are taken to train
replacements for technical men going into the armed services."[33] A subcommittee appointed by Bush on "scientific talent" included Watson Davis and astronomer Harlow Shapley, and it studied the Science Service projects Science
Clubs of America and Science Talent Search as examples of programs that
could innovatively circumvent school curricula so as to reach young potential scientists. In the end, the report omitted mention of the Science Talent
Search—a bit of neglect that angered Shapley.[34]

The Korean War meant a new expansion of scientific manpower fears
as the military and the contractors who served it struggled to fill the nation's
military needs. The National Manpower Council, formed in 1951, held conferences and issued reports throughout the 1950s on topics relating to the workforce, including education, public policy, and democracy.[35] As "manpower"
recurred again and again as a topic of concern in the 1940s and 1950s, science
education took a central place in the conflict over the future of progressive
education that defined the educational scene during the postwar era.[36] As life-
adjustment curricula moved into schools, science and engineering education
were depicted as victims of an education system devoid of rigor and rationality. Critics such as U.S. Navy admiral Hyman Rickover attacked schools' in-
ability to "save" smart students who wanted to undertake the hard intellectual
work of science research.[37] Partially as a result of this outrage, scientists suc-

ceeded in garnering public funding for scientist-led curriculum reform efforts in the 1960s.[38]

Changes in the scientific profession during the postwar era, as well as a cultural climate that scientists perceived as unstable and inhospitable, led to a far greater interest within the profession in the experience of "science-minded" youth. Science Service's work with the Science Talent Search, which often drew upon the resources of established scientists as lecturers or mentors, participated in these debates within the scientific profession about education and culture; in its efforts to encourage youth, Science Service attempted to create a new youth culture, in which curiosity would be rewarded.

SCIENCE SERVICE'S INTENTIONS AND GOALS

The Science Talent Search was founded and conducted by Science Service, an organization that evinced a public-spirited belief in the possibilities of social improvement through science—a characteristic attitude of the Progressive Era and the interwar period. As an organization steered by scientists and journalists and staffed by journalists with scientific educations, Science Service had a strong belief in the merits of extracurricular scientific education, as well as a commitment to science's public visibility. Science, the Service's founding ideology held, was for everybody, and scientific knowledge and curiosity was a key aspect of citizenship in a democratic society.

The Service originated in the early years of the century, when E. W. Scripps, the newspaper magnate, began to patronize the work of William E. Ritter, a zoologist and marine biologist. The two men formed a lasting friendship based on their mutual interest in the question of public education and its relationship to democratic ideals. Ritter believed in the benefits of public knowledge of science, which he defined both as familiarity with the facts born of the scientific enterprise and as ability to use the structures of thought inherent to scientific process. Scripps, a strong-minded, self-professed "damned old crank," wanted to fund reporting about "all things of human concern," a Scripps-defined category of journalism he believed would strengthen democracy and promote civic reform.[39] In 1921, Science Service was launched as a not-for-profit corporation, with enough financial support from Scripps to ensure that it would not find itself, as its first editor, Edwin E. Slosson, termed it, "under the control of any clique, class or commercial interest . . . [or] the organ of any one association."[40] Slosson, who had a Ph.D. in chemistry, was a committed science popularizer and the author of the popular books *Creative Chemistry* (1919) and *Easy Lessons in Einstein* (1920).[41]

The main remit of Science Service in its early years was to produce and to place accurate popular scientific writing in newspapers around the United States. At the beginning of its efforts, it sent a mailing once a week to participating newspapers, consisting of packaged stories with a "By Science Service" byline.[42] Beginning in 1922, the Service produced a weekly magazine called *Science News-Letter*, which was meant to aggregate the Service's coverage in one periodical for the use of individuals and schools.[43] Science Service also produced a radio show, eventually called *Adventures in Science*, which mixed recent headlines of scientific interest with interviews of scientists.

After Slosson's death in 1929, Watson Davis, a civil engineer-turned-journalist who had been working at the Service as managing editor since 1923, was appointed director. He retained this position from 1933 to 1966 and was a key figure in the organization, promotion, and longevity of the Science Talent Search. During his tenure at Science Service, Davis wrote and lectured widely on such topics as the nature of scientific progress, the manpower crisis, and the meaning of "science popularization," and he was in demand as an expert on the topic of science's presence in public life.[44] Despite a prevailing postwar trend toward coverage that would promote "science appreciation," rather than an understanding of scientific method or a deeper engagement with the questions that provoke research, Davis remained ideologically committed to the older Progressive idea of widespread science understanding as a way to strengthen the democratic process.[45] In 1948, for example, he wrote, "If the great mass of the people, through accurate and interesting accounts of the successes and failures of science, can glimpse and understand that essence of science, its trying, testing, and trying again, if they build their own convictions that this is a good, sensible, successful, and useful method, then there is hope that they will apply it more widely to everyday life, to our human relations, to running our businesses, to our governments, to everything that we do."[46] Besides retaining these humanistic commitments, Davis was also an internationalist, and under his leadership Science Service cooperated with the State Department, offering assistance to its book translation program; worked with the United Nations Educational, Scientific and Cultural Organization (UNESCO) to assist science popularization efforts in other countries; and published its reports in cities abroad.[47]

Partially inspired by the work done in the 1930s by the American Institute and the American Association for the Advancement of Science, Davis began conceiving Science Service's multiple programs for young people in the late 1930s and launched many more in the 1940s. During the war, Davis and Science Service tried to augment the more generalized science education that was the only thing available in many schools with extracurricular activities that would encourage children in science clubs to contribute to the war effort and civilian defense and to consider training for a science career to strengthen America in any future wars. Science Service took over the network of science clubs sponsored by the American Institute in 1941, quickly expanding its reach. The number of science clubs ballooned during the war, as students tested water, built model airplanes for airplane spotters, studied first aid, and discussed "scientific" strategies for keeping morale up on the home front.[48]

After the war, amid conversations about manpower, Davis and the Service invested heavily in youth programs. In explaining this move, Davis pointed to an increased degree of scientific progress as a reason why young people needed more and more help in their attempts to know science. In a 1948 speech to a group of science teachers in Cleveland, Ohio, Davis used the increasingly out-of-fashion language of recapitulation theory to drive home the urgency of teaching science in the postwar era: "Today it is recognized that science education must be accelerated if growing boys and girls are to recapitulate the scientific history of the human race in the few years between entering school and getting to or through college."[49] Davis also saw support of what he called "the youth interest in science" as an integral part of "science diffusion."[50] In that 1948 speech to Cleveland's teachers, Davis wrote, "If Johnny and Mary have such fun with science in their science clubs, there may be some hope that older folk will tumble to the fact, as thousands upon thousands have already done, that science is a good hobby."[51] The young scientist might serve as ambassador from the modern world to fusty and confused parents, converting more citizens to the cause of science.

Under Davis's direction, in 1940 Science Service began a series of mail-order materials aimed at young people, titled "THINGS of Science," in which teachers or students received a monthly sample in a little blue box. The materials could be natural or man-made and arrived along with explanatory text and suggested experiments.[52] Other Science Service publications were specifically targeted to young scientists, such as *Scientific Instruments You Can Make* (by Helen M. Davis, Watson Davis's wife), *Science Exhibits* (also by Helen Davis), and *Thousands of Science Projects* (by Margaret E. Patterson and Joseph H. Karus). These drew upon the bank of projects produced for STS and National

Science Fairs and listed titles of past projects, along with suggestions for shaping project goals and making equipment. The student (or teacher) could also order a box of slides of National Science Fair exhibits, along with related commentary.[53]

The Science Clubs of America, the umbrella organization created by the Service in 1941 to facilitate existing clubs founded by science teachers that had previously been administered by the American Institute with the support of Westinghouse, was probably the Science Service youth effort that had the largest membership.[54] In 1958, Science Service claimed that 400,000 young people showed their exhibits at Service-assisted fairs annually.[55] In the *Science News-Letter*, short reports on the doings of the science clubs offered small portraits of group interests and efforts. Using such names as the "Atomettes" (Newtown Square, Pennsylvania), "The Curiosity Club" (Normandy, Florida), "Explorer's Society" (Alexandria, Virginia), and the "Riley High Mad Scientists" (South Bend, Indiana), the clubs reported taking field trips to industrial sites and farms, giving "magic shows" in cafeterias, and creating small museums in school lobbies.[56]

The idea for the Science Talent Search came about after Davis met G. Edward Pendray, an employee of Westinghouse Electric Corporation, when both were in the process of planning for the 1939 World's Fair.[57] Westinghouse had hired Pendray, a science journalist, author of science fiction, and president of the American Rocket Society, as an advertising and public relations assistant in 1936, offering him a large budget and creative latitude. Westinghouse's chairman of the board, Andrew W. Robertson, believed that the company needed to enhance its public reputation as forward looking. Pendray's idea to bury a time capsule at the 1939–40 World's Fair was one example of the kind of publicity stunt he offered in response to this mandate.[58] During the fair, several thousand boys and girls had the chance to show their science projects in a building sponsored by Westinghouse.[59] The idea for the Science Talent Search emerged from Davis and Pendray's collaboration on this event.

The difference between the National Science Fair and the Science Talent Search was the difference between a democratic, or inclusive, and meritocratic, or selective, vision of science education.[60] This contrast, which can be seen solidifying in the STS's promotional efforts during the 1950s, represented Science Service's dual ambitions in its youth programs: to cultivate scientific feeling in the masses, and to identify and support shining stars of research.[61] Davis and Science Service described the Science Clubs of America and the National Science Fair as the "grassroots," a democratic testing ground from which scientists with "aptitude" might emerge. Many Science Talent Search finalists were participants in Science Clubs of America, while some were previous finalists or winners at the National Science Fair.[62] In the meantime, through the nationwide network of science clubs, "millions . . . who do not and should not become scientists and engineers, experience science as a hobby, to their personal benefit and to the enrichment of our national policy."[63] In order to do that, Westinghouse and Science Service needed to define the ineffable spark that animated the "embryo scientist."

THE TALENT SEARCH AS CONTESTED EXPERIMENT

Presenting a plaque to Marion Cecile Joswick, one of the 1945 STS finalists, at Brooklyn Manual Training High School, Science Service employee Margaret Patterson described Joswick as the kind of person of which "our world has great need": "People who will try new things; people who have faith in themselves to persevere long enough to accomplish them in spite of any difficulties; people who are willing after one 'mission accomplished' to go on to another more challenging." Science Service, Patterson said, was "firmly convinced that such people exist" and that it was possible to find them while they were still young and "help them become more completely the people this world sorely needs."[64] By a bit later in the 1940s, Science Service spoke of children with "science talent," who had managed, despite the educational and cultural odds, to make it to their senior year having maintained their interest in science, as national resources to be protected at all costs. "Our great problem is to see that these potential scientists of tomorrow have a chance to show their worth," Watson Davis told an audience of science teachers in 1948.[65] In arguing that the Science Talent Search could find the kinds of students who had already proven themselves to have "scientist" qualities, Science Service was also making arguments about what it meant to be a scientist and the place of independent and challenging "work" in a young person's life.

In structuring the elements that went into the Talent Search, the adults running the Search in the postwar decades were themselves curious. They wanted to know how—in the words of the psychologist Harold Edgerton, who served as a judge and authored the Search's science examination—these students had "got that way." The Talent Search was, to them, a grand experiment, with the successful finalists serving as raw material. The adults doing the searching saw themselves as scientists, as Davis wrote in 1950 about the selection process: "There is promise that traits [of science talent] can be discovered and analyzed much as the chemist assays promising ore for its chemical elements."[66] In his introduction to the STS's "alumni" magazine, the *Science Talent Searchlight*, Edgerton told the students that they were "guinea pigs"; "you are being watched carefully: not to pry into your life as an individual, but to find out more about scientists and how they grow."[67] The pun of the magazine's name was meant to emphasize the degree to which the designation of students as STS winners would put them under scrutiny. Perhaps because these students were themselves scientists, Science Talent Search personnel felt no qualms in informing them of their position as lab rats and, in fact, tried to incite their scientific spirit in appealing to them to respond to surveys and contribute their experiences to the database of STS knowledge.

The final actor in the metaphor of "STS as experiment" was Westinghouse, framed as a patron of forward-thinking research in its underwriting of the Search. In supporting the Search, Davis argued, Westinghouse was committing to the future, in the same way as it might by supporting basic research done by adult scientists. The current work of young scientists might seem unfocused or exploratory, but its future potential was unlimited: "This is an excellent example of the support of 'pure' research by industry, because the chance of Westinghouse adding to its staff any of these talented young people is probably in about the same ratio as the likelihood of using some of the fundamental researches in its laboratory."[68] The equation—childhood is the "pure research" phase of life—compliments both young people and scientists, lauding the purity and possibility of their work. The metaphor also compliments Westinghouse, an entity foresighted (and deep-pocketed) enough to see the worthiness of investing in inquiry without an immediate yield.[69]

This "science experiment" of the STS was rooted in the wartime and postwar ascendancy of the field of psychology, which was gaining influence in government bureaucracies, universities, and the public consciousness.[70] Where academics and policymakers involved in planning for and analyzing social life might have previously tried to look to structures or institutions as explanatory factors, the psychology of individuals and groups now took precedence.[71] From 1930 to 1950, the creation of New Deal agencies (largely staffed by social scientists), the labor involved in mobilizing the American armed forces and crafting home front policies for World War II, and the advent of a cold war that relied heavily on psychological analysis combined to create a climate in which social science came to greater prominence.[72] During the Cold War, social scientists could be found working for all branches of the military, intelligence agencies, departments of the government, civilian advisory groups, and private foundations.[73]

The "experiment" of the Science Talent Search was part of a larger postwar push for research into the scientific vocation, especially its relationship to the nature of human creativity. The relationship between individual human hopes and desires and the normative demands of larger society was a quintessential preoccupation of postwar psychology. Interest in creativity was a way to formulate a response to the perceived threats of conformity and authoritarianism.[74] As postwar psychology's new interest in the individual and creative in human thinking expanded, the question of how scientists think and work was central to the field's inquiries. Humanistic cognitive researchers trying to move the theory of how people think beyond a simple stimulus-response behaviorist model looked to scientific thinking—much like the kind of work they themselves were doing—as a model subject.[75] This observation must be kept in mind when evaluating the arguments made by scientists about the conditions of childhoods of scientists. The reflexive nature of this project meant that many investigators and outside observers from the broader scientific community felt quite strongly about the findings and expressed themselves accordingly.

Concerns about enhancing human creativity often intersected with the debate over parent-child relationships that had become more prominent as the postwar baby boom reached its peak. By the 1950s, discussions around permissiveness had become dominant in the discourse around parenting, with books and magazines intended for parental consumption addressing every aspect of the new approach to parent-child relationships.[76] Popular culture intended for children, including television programs, films, records, and literature, changed in response to this new climate, while manufacturers of toys and children's furnishings incorporated the appeal of "developing creativity" into their advertising.[77] Although the education reforms of the 1950s understood themselves to be in revolt against overly child-centered pedagogical theories, there were continuities between the ideologies of science promotion and permissive child-rearing: both believed in unleashing the child's curiosity and imagination. The young scientist would have to embody an ideal (and sometimes paradoxical) combination of creativity and self-discipline.

Science Service and Westinghouse often used the word "creative" to describe the type of young scientist they were seeking, conflating the term "creative scientist" with the type of scientist who would produce original basic research. Even on the occasion of the first STS in 1942, psychologists Harold Edgerton and Steuart Henderson Britt were employing the term "creative" to describe the ideal STS finalist. By the 1950s, the question of "creativity" had become a common point of discussion at the Science Talent Institute, which now took for granted the idea that it was trying to encourage future researchers of the highest (creative) caliber.

A range of psychologists, theorists of education, and sociologists applied themselves to the project of understanding the scientific vocation in the 1950s, and many of their projects looked at factors in the childhood of scientists that might enhance or discourage creativity and shape a future vocation in science. The University of Utah hosted seven conferences between 1955 and 1971 that attracted many of the most prominent researchers within the creativity movement in psychology; the first three of these were devoted to "identification of scientific talent," while subsequent conferences addressed the concept of creativity more broadly, asking how creative people worked, wondering what their backgrounds were, and investigating the function of creativity in particular occupations and fields.[78] The National Science Foundation funded the conferences on the identification of scientific creativity.[79] Psychologist Anne Roe, an attendee and member of the steering committee of the University of Utah conferences, published perhaps the most visible work of analysis of scientists and their vocational choices, the 1953 trade nonfiction book *The Making of a Scientist*. This work surveyed sixty-four male biologists, physicists, and social scientists, all judged to have been professionally successful, using interviews and projective personality and intelligence tests.[80] Other studies that addressed the childhood and youth of scientists included the 1955 volume by Paul Brandwein, a science teacher at Forest Hills High School, in Queens (a large comprehensive high school that sent many students to the STS); the NSF-funded survey of science students "lost to college," by Columbia assistant dean Charles Cole, published in 1956; and a study of undergraduates at liberal arts colleges, published in 1952.[81] In addition, in his work on the "unsolved problems of the scientific career," Lawrence Kubie, a psychiatrist at Yale, who had corresponded with Roe while she was in the process of compiling her work, proposed several hypotheses about the psychology and lives of young scientists that were mostly based on his own experience and his knowledge of other scientists' lives.[82]

In their reports, researchers evaluated the place of family, school, and peer culture in the young person's growth as a scientist and a creative person. The researchers looking at the place of family life in the intellectual biographies of young scientists used this evidence to make arguments that the new permissive parenting would eventually yield intellectual stimulation. Several researchers mentioned *The Authoritarian Personality* (1950), an influential text for creativity researchers, in which Theodor W. Adorno and his coauthors theorized that excessive discipline in the home created rigidity and dogmatism.[83] Those investigating the roots of scientific creativity thought that a permissive, encouraging atmosphere might allow for flexibility and innovation in thought. For example, Maury H. Chorness, of the Air Force Personnel and Training Research Center, reported to the Utah conference that predictors of family environments promoting creativity included the way that parents reacted when their children used household items as toys; the "level of interest or irritability manifested by parents in hobby items or toys found underfoot"; and the degree to which parents were good at thinking up things for children to do during inclement weather.[84]

The dangers of authoritarianism, evident in the family setting, extended to the classroom. Paul Brandwein thought that teachers who created autocratic laboratory environments tended to lose students; "when a permissive (not coercive, autocratic, or laissez-faire) attitude prevailed, there was a noticeable growth in the ability of youngsters to work effectively."[85] Authority, Brandwein argued, was poison to the budding scientist, who needed to cultivate an attitude he called "questing." "The general acceptance of authority in a given field of scholarship without question and without ascertaining the reliability and validity of the authority is not characteristic of questing," Brandwein wrote. "The belief that all is well in this best of all possible worlds is not questing. . . . Questing arises in a dissatisfaction. Questing . . . results in curiosity."[86] Roe described what she saw as a climate in schools that squelched intellectual curiosity; many of the scientifically able students might have conflicts with authority figures, as "many of them are brighter than their teachers, and can think up a lot of things that are very difficult for teachers to cope with."[87] Psychologist J. W. Getzels and educational theorist P. W. Jackson had warned about the fate of "divergent" students: "Divergent fantasy is often called 'rebellious' rather than *germinal*; unconventional career choice is often labeled 'unrealistic' rather than *courageous*."[88]

If, in the findings of these researchers, young scientists could be hurt or helped by parents and teachers, their special hell was their peers. "Social integration or isolation" was a problem for many of Roe's respondents. Roe found that both biologists and physical scientists exhibited this pattern: "the rather shy boy, sometimes with intense special interests, usually intellectual or mechanical, who plays with one or two like-minded companions rather than with a gang, and who does not start dating until well on into college years."[89] Roe speculated that many of the respondents had maintained their childhood curiosity and developed it in intellectual directions *because* they had failed to develop socially or physically; "it is evident that a boy who cannot, for some reason (e.g., physical disability, or an immediately older brother) compete effectively in sports can gain at least some status by surpassing the other boys in school work."[90] Roe worried that this strategy might cause gender trouble; she thought scientific boys were sensitive and worried that they might "be derogated by athletic boys, who are likely to be the big shots in high school." Meanwhile, "many intelligent girls never learn to reconcile their type of intelligence and their femininity."[91]

The concern on the part of these scientists for their younger counterparts sometimes departed from objectivity and veered into bitter reflection. Near the end of his paper on the personalities of scientific researchers, psychologist Raymond B. Cattell argued that the paucity of "pure research" coming from American science was due to the "cult of the extrovert" in American schools. Admitting that he was "seasoning the dish with definite personal value judgments," Cattell wrote, "Whereas the schools for at least two generations have cherished the ideal of the extrovert, almost as if it were synonymous with mental health, the evidence is overwhelming that the creative person is an introvert." This "cult" came along with "worship of conformity, fads, and fashions," "low regard for intellectual activity," and a "preference for the witty over the wise, the casual over the exact, and the verbal, emotional, and superficial over the thoughtful, objective, and penetrating."[92]

In their experimental design, observations, and hypotheses, researchers looking at children's interest in science agreed that young scientists faced an inhospitable social milieu. If Mead and Metraux had found in their evidence that high schoolers disdained adult scientists, here was proof that the young scientists who were their peers were themselves the targets of scorn. In the 1950s, Science Service began to see the Science Talent Search as one method of shifting this culture; in its selection of young finalists, it hoped to find the "well-rounded" student who also happened to be interested in science. Displaying these specimens to the world would change people's minds about what it meant to be a scientist. This strategy was not without its naysayers.

THE TALENT SEARCH AND ITS DISCONTENTS

During the 1940s and 1950s, psychologists Steuart Henderson Britt and Harold Edgerton held a large amount of influence in the selection of the Science Talent Search finalists and winners. Britt and Edgerton wrote the Science Aptitude Examination, which was a major component of the STS selection process, and served as two members of the committee of judges who evaluated the finalists. (During the early years of the STS, astronomer Harlow Shapley was the third member of the committee; beginning in the 1950s, his place was taken by Rex Buxton, a psychiatrist from New York City.)[93] Both Ph.D.'s in psychology, Britt and Edgerton were shaped by their experiences working for the government during World War II; both were also employed in the private sector later in their lives, consulting with businesses about consumer and employee behavior. Britt enlisted in the navy during the war, employed to select and train personnel; after the war, he worked in advertising and as a marketing consultant.[94] Edgerton, who identified himself as an "industrial psychologist," worked for universities, for the U.S. Employment Service, and for a range of corporations in his capacity as a consultant.[95] Britt and Edgerton used the data they garnered from their work with the STS to publish papers on the question of scientific vocation.[96] Other psychologists and educational theorists also used the lives of the STS winners and finalists as raw material: Robert Mac-Curdy's 1954 dissertation, for example, surveyed the STS winners and finalists of 1952 and 1953.[97] Edgerton's student Ralph David Norman wrote a dissertation surveying the winners from 1941 to 1947, asking about their membership in honorary societies, their grades in college, publications, patents, and proficiency in using specialized devices or apparatus.[98]

The process of selection of the forty "winners" each year was composed of what Britt and Edgerton called "hurdles." The order of these "hurdles" changed from year to year but included submission of a completed Science Aptitude Examination, the Personal Data Blank (which asked the teachers "in the best position to judge the fitness of the student for the further study of science" to assess the student in categories such as "attitude-purpose-ambition," "scientific attitude," "work habits," "resourcefulness," and "social skills"), an essay, and complete school transcripts.[99] In 1966, Edgerton reported that about 2,000 to 4,000 high school seniors entered the contest yearly; 300 were listed as "honorable mentions," and 40 won a "Washington trip" and were designated "winners," then undergoing an interview with the Board of Judges.[100] This interview was the basis on which the judges picked scholarship winners.

Each student was interviewed separately and asked standardized questions meant to assess the strength of the student's preparation for a science career, as well as the student's desire ("drive") to have such a career.[101]

Even as some researchers saw the STS data as a great source of information about the scientific vocation, the project faced criticism from within the scientific community. In 1945, Paul Brandwein wrote to the letters department of *Science* with a one such critique: "The 'Science' Talent Search is in its fourth year. As a teacher of science . . . the writer has regularly brought it to the attention of all science students, has complied with the rules of the contest and has sent the papers of the contestants to the examination committee. During these years, the writer has shared with others the feeling that this may not be a science talent search." Overly dependent on teacher recommendations and facility with the written word, the STS process selected for scholastic excellence, rather than "science aptitude." Brandwein criticized the certainty with which the sponsors of the STS had promoted the outcome as a definitive determination of what it meant to have "science talent." He asked, "Is it possible that students who can not succeed in the written examination and who were successful in the other parts, if given the publicity and opportunities afforded the winners, might make equally good scientists?" As it stood, he said, "the present Science Talent Search could well be called 'Scholarships For Good Students with Present Interests in Science.'"[102]

Despite Science Service's continual emphasis on the "creativity" of the science students they sought, the Science Aptitude Examination itself included no questions that would test the students on the novelty of their thoughts. Edgerton used the word "aptitude" to describe the examination, aligning his work with the postwar trend toward use of standardized aptitude testing as a way of fairly assessing the merit of college applicants. So, for example, Edgerton wrote in the Science Service publication *Science News-Letter* in 1942 that the test was designed not to put "a heavy premium on previous knowledge of science" but rather to "select those who have the aptitude to study science in colleges and universities."[103] Edgerton, who composed tests designed to assess aptitudes in various populations throughout his career, created twenty-nine versions of the examination; because the test was released to the public after being administered each year, every year's test needed to be new. Edgerton wrote, in what critics like Brandwein might perceive as a telling admission, "Basically it was an academic aptitude test dressed in science clothing."[104]

After the first year, the Science Aptitude Examination's format settled into a three-part structure. The first section had fifty multiple-choice questions. Despite Henderson and Britt's contention that the test would measure *aptitude*, rather than mere content knowledge, there can be no mistaking the fact that these questions tested the respondent's mastery of vocabulary, taxonomy, scientific equipment, and the history of science.[105] The second section, with fifty questions, provided paragraphs drawn from scientific literature, designed to be unfamiliar to the student, which were spread across disciplines. These paragraphs were meant to provoke students to use discrimination and selection in processing new types of knowledge.[106] For each paragraph, about half the multiple-choice questions could potentially be answered based on reading comprehension; some asked the student to pick an inference that was fair to make based on the information contained in the paragraph, and some asked for an application of given mathematical formulas to another situation. The third section, also comprising about forty or fifty questions, was meant to test knowledge attained through means other than reading. Some of these questions were designed to measure mechanical ability, as in the diagram of a toy steam engine, with multiple-choice questions about the action of the mechanism.[107] Science Service reported that the scores ranged from 10 to 15 points (with a single point being awarded for each of 140 to 150 correct questions) to 110 to 120. Boys, on average, scored 12 points better than girls.[108]

The tests, published in the *Science News-Letter*, provoked criticism from several quarters, most notably from working scientists. In 1947, Frank Jewett, physicist, past president of Bell Labs, and then head of the National Academy of Sciences, sent Watson Davis a humorous letter about his attempt to take the Science Aptitude Examination. "Am horrified to find that . . . I have no aptitude whatever for science—think I'd better carve out a career in the butter and egg business!!" he wrote. Jewett thought, "If promise of capacity as a research man is a goal I suspect the boy would pass high in two or three sectors and fail miserably in all else. If he did well in a variety of sectors I should guess his career in science would be something where a walking encyclopedia was sought. . . . Possibly I'm too flippant in my ignorance but this sort of test would have been Greek to me in the days when I was picking out men for Bell Tel. Lab."[109] Davis took this criticism seriously, writing to his assistant in a note marked "URGENT": "Give me reprints that explain that if you do not make 100 on the STS test you are not a dumb-bell."[110] Replying to Jewett, he was careful to explain, "No one is supposed to make a perfect score on the Science Aptitude Examination" and that the test "is only part of the selection technique." Most of all, he wrote, "you have, of course, put your finger on the essential question which has been uppermost in our minds: can STS select creative scientists?" Davis added, "Our attempt is only six years old and we are confident that we have selected some who will be good scientists, judged by what they have done in undergraduate and graduate work."[111]

If Jewett thought that the test favored "walking encyclopedias," others believed that the overall process was overly kind to the well-rounded kid, as opposed to the scientific obsessive. Banesh Hoffmann, a physicist employed at Queens College, responded in the *American Scientist* to an earlier article in which Edgerton and Britt had described the "hurdles" of the selection process. Hoffmann, who in 1956 confronted the Educational Testing Service with a strong critique of the SAT, was later to write a wholehearted manifesto against the growth of standardized tests, *The Tyranny of Testing*. His major critique of the SAT and other tests was that they confounded the more intelligent students, who would find them confusing and poorly constructed.[112] Hoffmann thought that the reading comprehension questions in the Science Aptitude Examination would not measure scientific ability and suggested that if the test had been administered to a non-self-selected group of students (as opposed to the self-identified "future scientists" applying to STS), "would not one expect to find the future patent attorneys, the future authors of first rate detective stories, and other literary persons with clear heads edging out many of the genuine scientists in direct competition?" Hoffmann thought that the test, and the whole process, favored "the polymath at the expense of the specialist" and described the successful student in the STS, cuttingly, as "very clever." However, he pointed out, "it is an important fact that many very clever people are not scientists, whereas some very great scientists are, on the whole, rather childish and do not sparkle in their more superficial mental processes."[113] Like Jewett, Hoffmann wanted to save some room in science for oddballs whose talents were not immediately legible.

The Personal Data Blank allows us to see what Britt, Edgerton, and Science Service thought "science talent" would look like from the vantage point of the teacher. In the *Science News-Letter*, Edgerton wrote, "While there has been a classic picture of the scientist as a 'lone wolf,' a modern version is an individual able to think for himself, to lead others, and to work cooperatively. A scientist must be a well-rounded human being."[114] Edgerton and Britt, like the "management man" seeking scientific personnel for the company lab who found himself so maligned in William Whyte's 1956 book *The Organization Man*, sought "well-rounded" scientists who could work collectively.[115] Edgerton and Britt asked teachers about students' "independence": the student should have a "purpose and a program," be a "self-starter," and should not "always follow 'the way it was done by others.'" They wondered about "reliability": the student should "attend to details, finish his work on time, stick to the task until it is finished, work steadily at an assigned job"; he should be able to be "trusted with money, property, and confidential information." They asked teachers to rate the student's rationality: "Is he objective about most situations or does he react emotionally?" At the same time as he possessed these attributes related to character, the student should also shine in the personality department, "have the ability to direct others, to gain the whole-hearted cooperation of students and associates, to make a favorable impression on persons he meets."[116]

The objections mounted to the selection process echo larger controversies over the evolution of the scientific profession. The Personal Data Blank bothered Hoffmann the most; he asked, "Cannot a great scientist be lazy, shy, uncooperative, and utterly irresponsible?" He posited: "A boy who spends all his time in the fields watching birds may neglect his other activities to such an extent as to appear lazy and good for nothing. He may be very shy and awkward in company. He may be uncooperative when it comes to the usual activities of his colleagues. And he may [be] truant and behave in other ways suggestive of serious irresponsibility. Yet all with a definite purpose in view." Hoffmann wanted to allow the child scientist latitude to be strange and to operate outside the typical organizations of childhood and youth: "Must a scientist be a socialite before he can achieve greatness — in science? . . . Must he be able to take on responsibility? After all, is it scientists we are trying to select, or boy scouts?" Hoffmann found this problem to be a serious one, given the STS's claim to prestige: "Should the system spread so as to become the only regular channel through which young scientific talent is officially recognized it would entail serious dangers for the future of American science."[117] The STS's branding, which brought it media coverage and presidential accolades, was a dangerous thing to critics who feared the power it might come to wield.

Edgerton and Britt were invited to read Hoffmann's article and respond to it in the same issue of *Science*, and they defended themselves on several points of procedure; more than anything, however, they responded to the "favors the polymath" critique by arguing "the interdependence of all scientific areas is becoming increasingly apparent . . . most of our outstanding mathematicians, physicists, engineers, and other scientists are 'well-rounded' human beings who are reasonably at home in areas of science in addition to their own particular field of specialization."[118] But most of all, in responding to criticism,

Edgerton and Britt pointed to results: the success of the finalists in universities and research labs proved, in their minds, that the STS was looking for the right kinds of young people.[119]

CINDERELLAS AT THE SCIENCE TALENT INSTITUTE

In 1945, Science Service's Margaret Patterson told the audience at Marion Joswick's high school that she was impressed by the likability of the first group of Science Talent Search finalists. "We were quite prepared when we invited the first 40 to Washington, to greet a group of 'brains' with very little else to make them attractive," she said, "but we were pleasantly surprised . . . to find them just about the nicest people you could hope to meet — and thoroughly human." The students were "well-rounded in their interests," though given to falling into discussions about science during any spare time in the program. They were "absentminded" (Patterson described a Brooklyn "lad" who continually lost his train tickets; "he is now making a splendid record at Harvard"); determined (she pointed to a West Coast boy who executed elaborate travel plans in order to carry out his dream of seeing New York City); and modest and self-deprecating (she indicated a boy refugee from Germany who ordered a whole broiled lobster and picked at the outer shell until asking the waiter for another, since there was "no meat" on that one; "he is now at Harvard and the research he does is as secret from us as the interior of the lobster was to him in those days").[120]

In its representation of the events of the Science Talent Institute and the students who attended, Science Service slipped between describing the STS finalists as resolutely "ordinary" — curious, self-motivated, and independent, to be sure, but also sociable and leaders among their peers — and calling them stunningly special. In the postwar period, the vocation of scientist was increasingly presented as quotidian. If, in the first half of the twentieth century, scientists had been perceived as special (priests, saints, or magicians); in the postwar period, some scientists worked to demystify their calling, describing it as a job like any other in the service of normalizing their profession. So, for example, Glenn Seaborg told young audiences (including, in 1961, those at the Science Talent Institute banquet): "There is plenty of room in scientific research for those who are not in the genius category."[121] The stakes of representing the scientific vocation in this manner were obvious: if the finalists of the Science Talent Search were "human," others could hope to repeat their accomplishments; because part of the rationale for conducting the search in the first place had to do with encouraging others to follow suit and with representing science as an accessible and fun vocation, then this normalization of the young scientist was instrumental in achieving these goals. However, Science Service promoters also had a stake in representing the finalists as exceptional. If the title of STS finalist was to retain its prestige, the students who found success would have to be impressive.

The finalists who went to Washington were treated to a whirlwind program of lectures, visits to laboratories, conversations with scientists, project exhibits, judges' interviews, and media appearances. The Science Talent Institute (STI) experience was, the testimony of STS finalists shows, both exhilarating and exhausting. Paul Cloke (a finalist in 1947) wrote humorously to the new STSers in the *Searchlight*: "If you can't find time to see Washington during the day, you can always see it after 11:30 pm; the STI has ended its session by then."[122] Unlike the National Science Fair, which moved from city to city each year, the Science Talent Institute was always held in Washington, D.C., a location that underscored the growing postwar ties between research science and the government and reinforced the expectation that STS finalists and winners would become part of the elite scientific establishment. The STI's choice of host city also meant that Science Service could bring the finalists together with politicians and show them through the halls of power; this allowed for many good photo opportunities. In a tradition that was to be repeated with every Science Talent Search in later years, the assembled group of finalists met and was photographed with the president (or, as with Henry Wallace, described as a "scientist-statesman," the Vice President). Many of the field trips on the program were to national laboratories, and program speakers were often drawn from Washington, D.C.'s roster of prominent scientists employed by the government to direct and promote the nation's scientific efforts. Speakers in 1952, for example, included Dr. M. H. Trytten, the director of the Office of Scientific Personnel at the National Research Council, and Dr. Alan T. Waterman, the director of the National Science Foundation.[123]

The Science Talent Institute programs regularly contained designated periods of time for the students to connect with older, accomplished scientists. The program always called this an opportunity for "winners to meet and engage in scientific conversation with leading scientists"—a meeting that would lead to another photo opportunity, this one less rigidly posed and more evocative of fellowship and intense conversations.[124] The last day of the five was devoted to "prime time," in which students could visit what Science Service called "foremost men and women of research" working around the Washington, D.C., area.[125] (In a cute twist of words, another press release called them "the country's leading grown-up scientists.")[126] In 1956, a Science Service press release said, finalist Daniel Ch'en, of Eugene, Oregon, "will discuss the Chapman-Stormen Current Ring and general problems with Dr. Harry Vestine of the Department of Terrestrial Magnetism, Carnegie Institute of Washington"; Thomas O'Brien, of Rochester, Minnesota, "will discuss problems of medicine in relation to space flight with Dr. Walton L. Jones of the U.S. Navy Bureau of Medicine and Surgery."[127]

The final group of people with whom the STSers were encouraged to hobnob was the other STSers. The program regularly listed appearances by former STS winners or finalists, who spoke to the "new STSers" about their experiences in college and the early phases of their careers. In 1952, for example, Mr. and Mrs. Richard Milburn, both of whom won STS prizes in 1945 and were later married, addressed the assembled finalists on "developing careers in science."[128] Reading the *Science Talent Searchlight*, it becomes evident that the STSers arrived at a group identity during their time in Washington. Alumni call each other "science kiddies" and "fellow embryos"; salutations like "Dear Gang" abound; letters are sprinkled with reports of other STSers bumped into around campus and entreaties to make it to the reunions held in New York, Chicago, and Boston around Christmas time. In 1948, Eugene F. Haugh (Class of 1947) enthused: "It surely is grand to belong to a gang like this; I'm eagerly looking forward to the Xmas reunion."[129] One finalist wrote to the Science Service staff: "I believe I learned more about how to make friends than ever before; at least I felt perfectly at home."[130]

If the networking activities of the Science Talent Institute were meant to impart a message to the contestants about their value and exceptionality, the words of the acclaimed scientists and members of the funding establishment who spoke at their awards dinner reinforced the message. Several emphasized the value the scientific profession placed on youth. Dr. Karl T. Compton, then president of MIT, gave the address at the awards dinner in 1944 and emphasized the importance of scientists to national defense, also making sure to tell students that in science, their youth was an asset. "A very large proportion of the greatest scientific discoveries have been made by men in their twenties or early thirties. . . . This may be because youth is more imaginative and less conservative than old age," he said. "Don't let yourselves be discouraged by observing that textbooks usually show pictures of great scientists as elderly men; this only means that their portraits were not painted until sometime after their great work was done."[131] At least one voice tried to temper the overwhelming approbation, as Harlow Shapley told the first class of winners at their banquet, not to get "vain or bumptious" because of this honor: "Most of these vain young scientists perish as scientists through becoming smothered in their own petty vanities and introspections." Older physicists during the 1950s and 1960s condemned the younger generation as spoiled by the external rewards that the profession now offered.[132] Shapley's remark seemed to make a similar generational criticism even before the postwar shifts in the scientific profession; however, he somewhat tempered the harshness of his critique when he added, "This distinction of being a winner in the Science Talent Search should be a source for sympathy, rather than for congratulations, because upon you heavy responsibility has been placed. You have no escape now from the necessity of hard work, persistent thinking, and sincerity in scientific activity. We expect great things of you."[133]

As the postwar years progressed, Science Service increasingly made sure that the finalists received exposure in the press, leveraging their youth as an unusual way to frame the discussion of scientific issues. In the 1951 issue of the Science Service publication *Chemistry*, the STS finalists were referred to as "The Famous Forty"; although this may have been wishful thinking on Science Service's part, the finalists did score a fair amount of media coverage through Science Service.[134] The STS finalists were interviewed on Science Service's *Adventures in Science* radio program each year from 1942 to 1958, discussing their projects and major scientific issues of the day, and often appeared in the organization's *Science News-Letter*.[135] Science Service often tried to score coverage in non–Science Service media by tying student projects to issues of contemporary concern. For example, in 1956 Science Service took a poll of the finalists asking how the United States could achieve "technological survival" in the face of the manpower shortage, then released the results of the poll to the press, hoping for coverage pegged to the ongoing worries about scientific manpower. The press release noted that students called for "high schools . . . to take as much pride in outfitting chemistry and physics labs as they do in outfitting their football teams." John H. Venable Jr. of Atlanta, Georgia, thought that "the winner of a science contest should be held in as high esteem as the school's star football player"—a not entirely neutral opinion, given his own status as finalist, but one that served the narrative for adult journalists ready to critique teenagers' peer culture.[136]

Photographs of postwar STS finalists, taken for publicity purposes, reinforce the image of a group of clean-cut kids who just happened to be superinterested in science. Photos of kids being examined by the judges, huddling

STS 1957 finalists sharing a milkshake. Courtesy of the Society for Science & the Public.

in intent conversation with scientists and engineers, sitting in on a legislative committee meeting, or meeting the president or vice president were supplemented by images of finalists cavorting in a swimming pool, sharing milkshakes, and participating in flirtatious co-ed snowball fights.[137] While Science Service used contestants' opinions on scientific matters or national policy as hooks to promote science coverage, the contestants' life histories were opportunities for reshaping public opinion about scientific professions. The "supplementary information" that Science Service gave newspapers—short life histories of the contestants, their projects, hobbies, and commitments—provided a chance for the Service to depict the contestants as healthy, well-adjusted people whose childhoods had been full of achievements both wondrous and ordinary. A 1955 STS press release described the scientist as a paradoxically omniscient and easygoing everyday citizen: "Today's scientist no longer fits the popular misconception as to his kind—a retiring, cloistered, head-in-the-clouds individual who shuns the company of all except the few who possess advanced know-how in his particular field of endeavor." To the contrary, the press release continued, "today's scientist knows about—and can talk about—

biochemistry and Beethoven, paleontology and politics, biology and baseball, dermatology and Democrats, radioactivity and Republicans."[138] Another press release from that year described the STS winners as impossible conglomerations of affinities: "a teen-age boy physicist-mathematician-chemist . . . who stars on his high school's varsity tennis team . . . a Colorado boy entomologist who excels in his chosen field, has a consuming interest in Shakespeare, and plays a hot guitar."[139]

As the ultimate token of normalcy, many Science Service press releases mentioned the athletic prowess of the finalists. This was particularly significant, because, as alluded to above, the older scientists who debated manpower issues in the postwar era regarded the degree of approbation afforded athletes in high schools with an envious eye. In one example, the National Science Foundation's founding head, Alan T. Waterman, remarked in his 1960 introduction to the reprint edition of *Science: The Endless Frontier*: "As a nation we still seem a long way from a universal understanding and appreciation for intellectual activity generally and probably will remain so until we attach roughly the same importance to academic achievement as we do, for example, to prowess in sports."[140] Science Service was not unaware of this oft-repeated argument. In a memorandum written to convince newspapers to support local science fairs, Davis quoted a cooperating editor: "It's a grand feeling to have one of my readers come up to me and say 'I'm so glad you are doing something for someone who isn't a half-back.'" Davis then went on to quickly qualify, in a parenthetical aside: "Don't misunderstand the reader, the editor, or us — for we are all in favor of football and many top science clubbers are good players."[141] Science Service often mentioned the height or body type of the male winners in their press releases, especially when it was impressive. In 1956, the release announcing the winners mentioned a "six-foot plus boy physicist" and a "six-foot, three-inch . . . boy physicist-engineer."[142] In 1960, the *Science News-Letter* described the "top young scientist of the year," Jerome G. Spitzner of St. James, Minnesota, as a "husky young farm-boy physicist" who is "hailed for his scholastic, scientific, and wrestling squad prowess."[143] By focusing on athletic feats and masculine physical appearance, Science Service claimed some of the approbation normally directed at athletes for its scientist winners, underlining its point that young scientists could be as "normal" as their peers; it also reinscribed a model of scientific achievement that was fundamentally masculine.[144]

While Science Service's official representation of its STS finalists adhered to the party line of the scientists being "leaders" among their peers —

abnormal only in their extreme normalcy—the Talent Search archives include a few documents that offer a messier look at the contestants' relationships to science, work, and leisure. Only one set of "biographical notes" on Science Talent Search finalists survives in the Science Service archives, from the Search of 1953. It is unclear who compiled these notes or what they were used for, though it seems fair to presume that they may have been used to create the supplemental information sheets that were distributed to newspapers. The raw content, however, contains much more insight into teachers' honest evaluations of their students than was included in the information sheets. In what seem to be excerpts from their responses to the Personal Data Blank, many of the teachers wrote that the students were *too* invested in science. Teachers reported that a male contestant from South Dakota "spends too much time at serious work—not enough on social activities," though they hastened to add that he was "not socially maladjusted in any respect." Another, from Florida, had improved in his social relationships recently: "A year ago [this student] was a very cocky lad, impressed by his own superior intelligence and disliked by many boys because of his 'chip-on-the-shoulder' attitude. He has mellowed and matured a great deal in the past year." A third, from New York State, was "lacking" socially. "Has too small a group of close friends, difficult to get to know—retiring and hesitant in conversation are his principal negative characteristics."

According to the teachers, the finalists' work habits also left something to be desired. A female finalist from Virginia was "slow in starting her work, but once she has begun her work it is not difficult to keep her interested in it." Another, from Indiana, "does not organize her time to best advantage. Sometimes is rushed at last minute." A male finalist from Oklahoma was "an overly self-critical person; this probably keeps him from doing more than he does." A boy from New Jersey "likes to be first to solve problems and therefore sometimes jumps to conclusions." Joanna Russ of New York (later to become an acclaimed author of feminist science fiction) was "such an enthusiast that she will undertake countless numbers of jobs before realizing she has started too many projects." None of the students whose teachers made these negative comments were top-ten finalists, with the exception of Russ; however, the fact that they made it over as many "hurdles" as they did—reaching the status of "trip winners"—means that Science Service had to allow for at least some introverted, impulsive, and unregulated behavior in their selection process.[145]

A final piece of evidence is available to show that the STS finalists worked more than would perhaps be considered "well rounded"—and that the discussion of this level of work was a core part of their culture. The alumni writing to each other in the *Science Talent Searchlight* spoke of work with a breezy commiseration; their descriptions of their personal levels of "busy" were elaborate and humorous. James B. Gibson (Class of 1946) reported to his fellow STSers on his work at the University of California, Berkeley: "I start studying . . . beating my cranium . . . against a stone wall . . . and my friends start yelling at me to 'Relax and be human!' HA! HA!! HA!!! GRRR!!!!"[146] Gibson dated his letter "?-?-'4?(?)," signifying his befuddled disconnection from the everyday world. Russell Johnson Jr. (also 1946) wrote: "As soon as I came back to school a reaction against pleasure set in and I have been working uncomfortably hard since. My schedule shouldn't be hard, but it makes me fair dizzy—as witness the original date on this letter—November 12—still visible thru the scratchings. There was no reason I shouldn't have finished the letter then—I just whirled away from it."[147]

MERIT AND INCLUSION

In their discussions of the methodology of the STS, Steuart Britt and Harold Edgerton were careful to emphasize what they saw as the meritocratic nature of their selection process. "*The names and geographical localities represented were completely unknown,*" they emphasized, "for this information had been blanked out so that identification was by serial number only. Also, no questions concerning either race or religion appeared in any of the forms used."[148] In several instances, winners of humble origins were singled out for press coverage; when the male 1945 winner, Edward M. Kosower, appeared on the radio show *Adventures in Science,* the press release was careful to mention that his father, who appeared on the program alongside his son, was a taxi driver in New York City.[149] A supplemental information sheet for 1957 finalist Warren Carleton Rauscher, of San Francisco, mentioned that he "credits his father, a meat processor, for triggering his interest in science."[150] The representation of STS finalists as hailing from all quarters suited Science Service's professed ideology, which held that science could be found everywhere and for all people. The idea of a young scientist being rewarded for an independent, joyful commitment to science, carried forward without access to expensive equipment or fancy school laboratories, meant that the lives of STS finalists could be reproducible, given a strong enough personal will; that they loved what they did; and that their rewards were entirely fair.

However, the family histories of the body of STS finalists reveal that, in many ways, their successes could be attributable to the social context into which they were born. In his study of the STS finalists, Robert MacCurdy commented that their families tended to be "stable, cultured, educated; enjoy economic advantages, [have] leisure time; they were often "democratic" and "permissive."[151] In 1959, the *Science News-Letter* admitted that 57.5 percent of the finalists had had scientists somewhere in their family background but added cheerfully, "Conversely, no scientists are recorded on 42.5 percent of the family trees."[152] In 1963 Science Service writer Shirley Moore wrote about the age that STS contestants first figured out that they liked science and analyzed the source of their interest, giving encouraging home influences ("home," "reading," "personal drive," "science equipment") a greater share of the credit than external factors ("school," "science clubs & fairs," "misc.").[153]

With the cooperation of Science Service and inspired by a Moore article, journalist Marianne Besser published a book in 1960 which surveyed both scientist parents and parents of STS finalists, asking many of the same questions: How do you raise a child who likes science? The answers were a primer in new-style family permissiveness: Parents should "share" and "guide" their children, rather than "pushing," "in the direction of the child's own interest." Mothers, especially, should try to join in their children's hobbies whenever possible. When choosing school courses, parents should encourage children to follow their interests, rather than trying to fit in by picking "snap" courses in order to get good grades; relatedly, "two-thirds of the mothers and fathers advised emphasis on research for the pure joy of the search, not mainly for honors and prizes."[154] Many of the mothers told the reporters that they needed to give up their own need to be a "good housekeeper" for the sake of the child's collections, and they advised other mothers to do the same; giving the child permission to be messy was a key part of the process of encouraging inquiry. Or, as Harold Edgerton told an interviewer in 1961, "If a very young child is fascinated by flowers, instead of ordering him away from mother's favorite petunias, let him dig one up, study the roots, feel it, smell it—satisfy his curiosity. Never belittle any scheme a child may dream up, no matter how wild it sounds."[155]

Both Moore and Besser found that STS parents were willing to provide their children with what Besser called "a lab of his own" in which to pursue scientific hobbies. Besser interviewed the mother of Eric Martz of Indiana, who told her about Eric's room, decorated with a snake cage, aquarium full of snapping turtles, a "disemboweled radio," and reference books; Besser argued, "Eric's room reflected his personality, interests, hobbies, and tastes—as it should." Besser acknowledged that space might be a consideration for some families, but thought that this limitation could be circumvented by giving a child a corner or a kitchen cabinet as a laboratory; however, the family of the contestant who grew up under these conditions, Philip Wagreich, eventually moved to a "house in the suburbs which had a basement where he could experiment"—and Philip's "science interests were a deciding factor in their move."[156] The centrality of children's needs in group decision making was a distinct feature of postwar middle-class family structure.

While the STS was nominally open to all races, African American faces were very uncommon in photographs documenting the STSers' Washington trips during the 1940s and 1950s. Moreover, in a 1949 report on the past year's statewide Science Talent Searches, which were conducted alongside the national STS, Margaret Patterson reported (without comment) that Alabama segregated the contests by race; while white winners received the generous prize of tuition and fees paid for four years at one of four Alabama universities, the "Negro" winner received no choice: a free ride at Tuskegee.[157] The celebrations were also separate; the white winners received an expenses-paid visit to the Alabama Junior Academy of Science Meeting, whereas the black winners convened at Tuskegee.[158] A few black contestants who made it to the national contest encountered difficulties in being able to fully participate. One of Nancy Durant Edmonds's fellow finalists from the year 1944 remembered that a group of contestants wanted to go to dinner one night but had to dine at the YWCA because Nancy, an African American, would not have been allowed in the city's segregated restaurants.[159]

The treatment of female contestants at the STS further complicated the picture of a disinterested meritocracy. Nominally, the STS was committed to shifting the culture so that women could join men in expanding the American scientific workforce. Edgerton and Britt wrote in 1944 that the contest's first three years revealed a significant difference in the scores of boys and girls on the aptitude examination.[160] The two used this evidence to defend the STS's decision to eliminate gender parity in the larger group of trip winners. (From the mid-1940s onward, the STS selected trip winners according to the ratio of boys to girls who submitted entries.) Edgerton and Britt argued that this difference between the sexes in scores indicated the need for better scientific training for American girls: "They [the girls' scores] are probably due . . . to environmental and cultural factors rather than to inherent biological differences."[161]

If such is the case, the STS was part of this culture. Promotional photos

STS 1954 finalists Marguerite Bloom (Burlant) and Robert Rodden
"look to the future." Courtesy of the Society for Science & the Public.

of STS girls depicted Nancy Milburn Stafford (1945) with her awards banquet
outfit and Carol Cartwright Hawkins (1955) sewing finishing touches on a gar-
ment in her hotel room. During the 1950s, the contestants staged a repeated
photo op that the STS called "Pointing to the Future," in which a boy and a
girl stood on the steps of the Capitol. The boy was always the one to point
(with one exception, the 1957 photo, which had three students in it, changing

STS 1960 female finalists look at the Hope Diamond displayed at the Smithsonian Natural History Museum. Courtesy of the Society for Science & the Public.

the dynamic). One 1960 photograph of three female contestants gazing at the Hope Diamond — an image of the girls' reflections in the glass over the gem — epitomized the STS's attempt to merge conventional femininity and scientific identity. The girls were at the nation's flagship natural history museum, but they were using their time there to ponder a spectacular symbol of female marital aspirations. The reflective composition of the image made the girls look wistful, emphasizing the intense nature of their desires.[162]

An interesting consequence of the manpower crisis of the 1950s was the increasing attention paid to the role of women in scientific and technical occupations. Although girls and women were identified as an untapped resource particularly valuable because they could not be drafted, many of those arguing for the expanded inclusion of women in the sciences were looking for girls to fill nonresearch jobs: lab technician, high school teacher.[163] The transformation of physics from a calling into a career, along with the increasing number of married male graduate students, meant that the environment became unwelcoming to female scientists-in-training.[164] Some called publicly for training funds to be reserved for men, given women's supposed propensity to drop

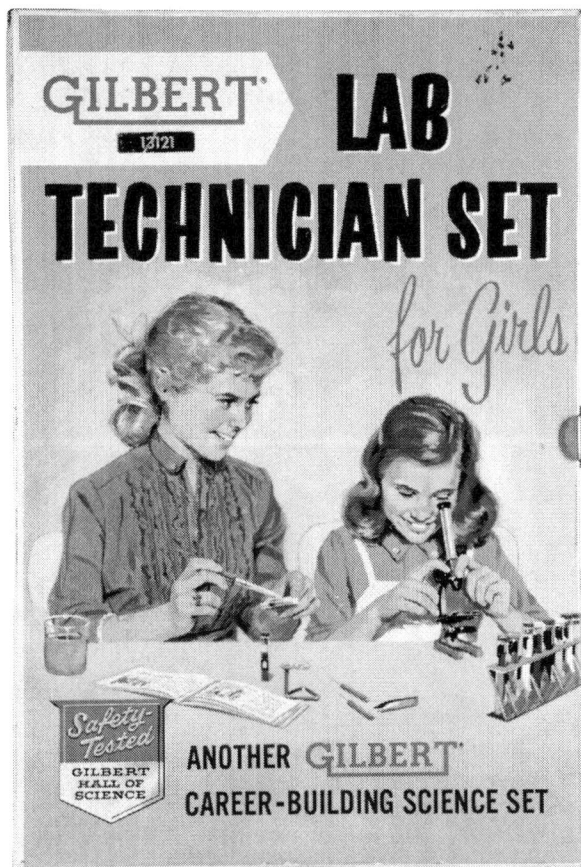

The cover of the Gilbert Lab Technician Set for Girls. Courtesy of the Chemical Heritage Foundation.

out of scientific careers in favor of homemaking. Advocacy groups for women in science, such as Sigma Delta Epsilon and the Society of Women Engineers, tried to lobby the government to support women in STEM, helped high school girls participate in science fairs, and provided career boosts for women looking to reenter the workforce after having children, but often fell back on feminine stereotypes in doing so, reassuring girls that they could be beautiful and desirable and pursue STEM at the same time.[165]

The Science Service's representations of young female scientists show the effort the Service was making to represent girl scientists as "normal," in the sense of postwar gender roles. Just as the Service tried to represent boy scientists as brawny and athletic, girls needed to be feminine and unprepossessing. The 1955 "girl winner," Kathleen A. Hable, of Loyal, Wisconsin, was described as "soft spoken, brown-eyed," and an "accomplished pianist." Hable was "poised, quiet, quick, witty" and described by her teachers as "do[ing] a lot of work without a lot of noise."[166] In 1960, the winner was Betty Lou Snarr, of Oklahoma City, described by the *Science News-Letter* as "a sparklingly feminine petite chemist"; the *News Letter* added, "The field of physical chemistry will be brightened a few years hence by the addition to its ranks of pretty, five-foot-two Betty Lou Snarr."[167] Girl STSers were portrayed as nurturing and family oriented. Girls' work with fellow students was portrayed as "helping," while boys were called "authorities" or "leaders."[168] The 1953 supplementary information indicated that Karen M. Spangehl of Phoenix "loves children to the extent of planning a career in pediatrics."[169] Merry Margolish, of New Rochelle, New York, a 1957 finalist, "has two aspirations for her future: raising a family and doing medical research because 'for a woman, a family should be as important as a career.'" In both press releases the year of her contest, the Service led with this quote.[170]

In early years of follow-up surveys assessing the careers of contestants, Edgerton and Britt, with their collaborator Ralph D. Norman, omitted the girls, as they were "comparatively fewer in number, and many . . . probably will not pursue active scientific careers."[171] Surveying the 1942 and 1943 contest participants and winners in 1966, Edgerton found that out of a sample of participants numbering 1,550, 80 percent of the female respondents to a questionnaire replied that they had left science (compared with 38 percent of males). Five out of eight of these female respondents had ("not unexpectedly," Edgerton said) become homemakers. On this subject, Edgerton found that the smaller sample of female winners interviewed gave him even more conclusive results. Out of 31 interviewees, twenty were married, and eleven were single. A third of the married women had continued to "build a professional satisfaction on top of home responsibilities," although these women "recognized that their husbands' job is of first importance . . . child bearing limited their professional work." Although those who were single reported occupational satisfaction, they "still found life incomplete and felt that marriage would be a desirable addition to or substitute for their job." Edgerton found no evidence that the single women in the sample had "remained single out of dedication to science; many expressed their desire for marriage, a possibility which becomes more remote as they advance in their professional competence and status." To Edgerton, these findings meant that there was "a need for realism in considering women as a manpower resource in scientific and technical areas, reexamining the social, economic, and education conditions under which they are a tappable resource."[172] For popular media, he put on a

cheerier meritocratic face, telling readers in response to the interview question, "What does a youngster need to become a scientist?" that, of 60,382
contestants who had passed the Science Search test to date, "14,007 of these
youngsters were girls, and that dynamites another myth—that females don't
make good scientists."[173]

Of the college students contributing to the "alumni" magazine *Science
Talent Searchlight*, many women wrote in about jobs or school; others, such as
Elizabeth Foster (STS 1943), reported having put schooling on hold in favor of
family. Foster wrote in 1948 about her upcoming marriage to a fellow graduate
student: "If I find that sweeping floors (or floor, depending on the extent of
our abode) and working at least part time (we're sharing expenses) does not
completely occupy my time, I may take courses toward a Ph.D."[174] Virginia
March Kline (STS 1943) reported on her life as a full-time housewife, which
left her feeling "isolated": "I'm tutoring organic chem. one afternoon a week
to help keep me from forgetting all I've learned in the long four years—hope
to go back to work one day."[175] In one issue of the *Searchlight*, a cartoon depicted a woman in a hospital bed, holding multiple babies, presumably in a
postpartum state; she says to a bedside visitor, "Thank goodness I finally have
something to report to the STSL."[176] The cartoon wryly normalized a situation familiar to the contest's young alumnae, depicting a young woman whose
onetime prolific scientific productivity had transformed, once and for all, into
human reproductivity.

CONCLUSION

In his 1960 novel *The Child Buyer*, journalist John Hersey depicted the sale of
a bright, oddball ten-year-old science aficionado to a corporation; the book,
written after Hersey had spent the postwar years advocating for better education of gifted children, was a scathing look at the adults surrounding the
young Barry Rudd, all of whom find different reasons (greed, blind admiration of authority, even scientific curiosity) to allow the sale. The treatment that
Barry will undergo at the hands of the corporation will shut him off from the
social world and the natural world, abstracting his brainpower until he is only
a cogitating machine. *The Child Buyer* asked readers to follow the implications
of the incessant calls for increases in manpower to their logical—and horrify-
ing—conclusion. What would a society be, if it refused to recognize a child's
right to a period of aimless and undirected play?[177]

Artifacts of children's popular science created in the prewar era, when children's scientific interest seemed common and easy, shifted focus in the postwar era to a career-oriented preparatory mindset. For example, the Brooklyn Children's Museum became "more didactic and task-oriented," hosted workshops and programs sponsored by the National Science Foundation, and acquired more and more books in chemistry, astronomy, and physics, while slowing down purchases of nature study books.[178] The Porter Chemical Company changed its motto in the mid-1950s from "Experimenter Today . . . Scientist Tomorrow" to "Porter Science Prepares Young America for World Leadership." Perhaps in imitation of the STS, Porter also created a national scholarship contest, which ran from the mid-1950s through the early 1960s.[179] Both examples show the influence on children's culture of public worries over America's continuing scientific and technological supremacy.

The Science Talent Search might seem, at first glance, to be a real-world version of the hunt for the gifted by John Hersey's United Lymphomblloid Corporation or to be invested solely in improving the output of young scientists with a view to "world leadership." But in examining its careful representation of its successful finalists as well-rounded, independent human beings, it becomes clear that Science Service was attempting not only to find future scientists but also to represent those promising students' talent in such a way as to create a new vision of the experience of being young in the United States. Whether or not the young scientists who arrived at the STS were actually as well rounded as the Service wanted them to be, they were not juvenile delinquents, obsessed with pop culture, or hopeless conformists.

In looking for the science-obsessed, those working for Science Service sought young people who, they believed, had—despite the culture that surrounded them—lived an archetypal childhood of an earlier time. In the postwar era, the adults looking to promote science play in young people's lives began to view their project as one of restoring prestige and interest to hobbies that (they believed) had formerly been common. In the space of a few decades, the science hobbyist had gone from appearing "modern" and advanced, as in the images of the "wireless boys" of the Brooklyn Children's Museum or the visions of experimentation on the tops of chemistry set boxes, to seeming vaguely forlorn and alone. From this time on, the young scientist was seen as a figure in need of protection—his curiosity fragile, his ego at risk.

Space Cadets and Rocket Boys

Policing the Masculinity of Scientific Enthusiasms

Homer Hickam, a West Virginian from a coal-mining town who started making and testing rockets in 1957, read an impressive slate of science-fiction authors as a young teenager: Jules Verne, Robert Heinlein, Isaac Asimov, A. E. Van Vogt, Arthur C. Clarke, Ray Bradbury. In his memoir of a Cold War scientific childhood, published in 1998, Hickam wrote that Verne was his gateway author: "I fell in love with Verne's books, filled as they were with not only great adventures but scientists and engineers who considered the acquisition of knowledge to be the greatest pursuit of mankind." Hickam's reading habits reaffirmed a growing interest in scientific experimentation, confirming its worth as a life pursuit, and putting him ahead of his elders when it came to scientific speculation. The feeling of being "ahead of the times" was addictive. When news of the Sputnik satellite broke in 1957, Hickam, then in high school, felt like his early interest in science fiction had primed him for the event. His mom woke him up, worried, asking him to explain what was happening. "All the science-fiction books and Dad's magazines I'd read over the years put me in good stead to answer," he wrote. "'It's a space satellite,' I explained. 'We were supposed to launch one this year, too. I can't believe the Russians beat us to it!'" Unlike the adults around him, Hickam remembers being more entranced than scared by the advent of Sputnik, buoyed by his total mania for all things related to rockets and space travel. "While I listened to the beeping" of the satellite as it passed over his town, he wrote, "I had this mental image of Russian high-school kids lifting the Sputnik and putting it in place on top of a big, sleek rocket. I envied them and wondered how it was they were so smart."[1]

By the time the panic of Sputnik hit, some small number of Americans—readers of science fiction and informal experimenters with rocketry—had been thinking about likely possibilities for interplanetary flight for a few decades.[2] Beginning in the late 1940s and early 1950s, a much wider swath of the public consumed space-themed popular culture. While interwar fans of space travel were adults as well as teenagers, consuming such niche products as science fiction magazines, 1950s space culture was mass market and mostly aimed at children and adolescents. As shows like *Captain Video* (1949–55), *Space Patrol* (1950–55), *Rod Brown of the Rocket Rangers* (1953–54), and *Tom Corbett, Space Cadet* (1950–55) dominated children's television in the first half of the 1950s, model rocketry grew in popularity, and manufacturers and publishers produced an array of space-themed toys, books, and clothing.[3]

This space-mad kids' stuff was not just for children; it also told adults a story about the direction of the culture as a whole. In the 1950s, the space cadet took his place alongside the butterfly collector and the basement experimenter in the adult pantheon of cute-science archetypes. Science fiction writer William Gibson noted of his own 1950s childhood: "The zeitgeist was chewy with space-flavored nuggets, morsels of futuristic design, precursors of a Tomorrow whose confident glow was visible beyond the horizon of all that was less wonderful, provided one had eyes to see it."[4] The rocket boy—a Baby Boomer encased in synthetic fabrics, wielding a futuristic toy weapon and populating his bedroom with diagrams of the solar system—was, as many a charmed adult was happy to point out, a native of the future. When Hickam and his crew began to test their rockets, a reporter from one of their town's papers came to all their test launches. He writes about their reaction to a successful test in purple prose that captures something of that feeling: "The boys race from their bunker and go running down their . . . firing range, the joy of youth and scientific interest playing across their delighted faces. Oh Rocket Boys, oh, Rocket Boys, how sweet thy missile's delight against the pale blue sky. A mile, a mile, they cry. We've flown a mile!"[5] The space kid was a walking, talking adorable reminder of Tomorrow, close to hand and easily observable for adults who might themselves feel too old to really understand the new interest in space.

In trend pieces, writers and illustrators for magazines and newspapers handled the space kid with loving attention. The lingo of the space cadet shows proved irresistible to writers looking for some humorous color. "The moon's attraction for a small boy is far greater, I maintain, than for any other object in our solar system," Jack Cluett wrote for *Women's Day* in 1953, comparing kids' waxing interest in outer space to what he thought was their waning attention to previous adventure fiction set in the Old West or the polar regions. "Space is the thing, now," Cluett wrote, "and if your house is as normal as mine, it resounds to the yells of *Blast off! What in the universe? You've lost your rockets, That's a lot of cosmic dust, Man the space locks, Stand by to reverse the gyroferge, Do you read me, Spartak?*, etc., etc., ad infinitum."[6]

Illustrators for general-interest magazines played with the visual contrast between the aesthetics of the space-kid shows and the actual midcentury world they lived in. The cover of *Collier's* on April 18, 1953, featured a boy in a "space helmet" holding a "blaster," looking adorably up at a butterfly caught in the bubble of his helmet — an image that humorously juxtaposed the imaginative life of the child with his actual earthly surroundings. On November 8, 1952, the cover of the *Saturday Evening Post* featured a boy and his mother boarding an airplane, with the boy clad in a domed space helmet. This space cadet is living in the future, even as he boards a spectacular technology his foolish parents once thought was impressive. The kids, the illustration tells the viewer, will always demand more; that's the way of the world, and it's what made the airplane possible in the first place.[7]

Most coverage of candy-colored space madness also tried to make a case for the ultimate social utility of children's newfound interest. In a few pieces for an issue of *Women's Day*, Cluett ran through the history of space travel in science fiction and scientific theory, telling mothers about the brief history of science fiction in the United States, while outlining what he presented as legitimizing instances in which science fiction had actually predicted new technologies before a given innovation manifested in the real world. Cluett also pointed to respectable scientists and engineers who were fans of — or worked on — space cadet television. "Willy Ley," he wrote, "is a walking encyclopedia on rockets, astronomy, ramjet engines, hydrazine, and reactor motors," as well as the technical consultant for *Tom Corbett, Space Cadet*.[8] Cluett quoted Frank Forrester of the Hayden Planetarium, an astronomer and science popularizer who was all for space fandom. Comics, television shows, and books, Forrester argued, were "all part and parcel of the new scientific age; and provided they eventually lead a youngster into the Planetarium, to Lynn Poole's brilliant and fascinating [nonfiction television show] *Johns Hopkins Science Review*, or into other paths of true scientific fact, as opposed to trivial fiction, I don't know but what they haven't *all* served a very important and significant purpose."[9]

The cover of *Collier's* magazine, April 18, 1953.

Women's Day ended its package with a feature showing moms how to make do-it-yourself helmets out of plastic bowls, sink drains, suction cups, and cambric.[10] Toy manufacturers saw an opportunity to convince parents that space-related interests could bring the family closer together. The 1960 brochure for Starmaster's science toys featured an image of a boy and his mom, with the mom sitting with skirts arranged prettily, gazing up at her son,

A page from a brochure for Starmaster Scientific Toys.
Courtesy of The Strong®, Rochester, New York.

who is examining a telescope with that sweet scientific joy on his face. "Make Space a Family Affair," the tagline ran.[11]

SPACE CADETS AS YOUNG-ADULT READERS

Young people often read the pulp magazines in which science fiction got its start in the United States in the 1920s and 1930s. Writers Isaac Asimov and Frederick Pohl have said that they discovered the genre at age ten, and Robert Heinlein began to read the pulps as a high school student.[12] In the interwar years, science fiction fandom was a group activity, much like the more educationally sanctioned, respectable chemistry and science clubs. Boys (and a few girls) used the letters columns of the pulp magazines to organize science and science fiction correspondence clubs. Groups such as the Edison Science Correspondence Club and the Scientifiction Association for Boys were explicitly designed to connect younger fans, in middle school and high school, with long-distance pen pals. The Scientifiction Association and other clubs pooled resources, contributing older issues of pulps to a group library and putting membership dues toward the purchase of books and magazines. In the 1930s, clubs began publishing fanzines; throughout the decade, fan clubs of adolescents and young men, clusters of which could meet in person in New York, Chicago, Philadelphia, and Los Angeles, argued about principles of science and of narrative, navigating a byzantine set of interpersonal alliances and divisions. Many of these fans were working class, urban, and sons of immigrant families. Their fandom was a world apart, largely unsupported by mainstream culture and by the nonfan adults in their lives.[13]

Everything changed in the late 1940s. While in the 1930s teachers spoke out against pulp magazines, lumping such science fiction magazines as *Amazing Stories* and *Astounding Stories* in with other cheaply printed publications said to be decaying the morals and minds of young people, in the 1950s science fiction was cautiously, and then heartily, embraced by forward-thinking librarians and educators.[14] Scribner's decision to print Heinlein's 1947 book *Rocket Ship Galileo*, and to publish it in hardcover, gave a first boost to the genre's credibility with these adult gatekeepers of children's reading.[15] Children's desires drove demand. In 1950, the Children's Book Council surveyed public and school librarians to get their perspectives on the need for children's books. Respondents asked for more science books for primary and middle-year readers; "There were many, many requests for science fiction at all age levels."[16] By the first and second years of the new decade, publishers of books for young people had geared up to meet that demand. The New York Public Library's recommendation list for teenage readers, Books for Young People — considered instrumental in the midcentury creation of the category of young-adult literature — included a "science fiction" section beginning in 1950 and a separate "Exploration of Space" section beginning in 1952.[17] After Heinlein entered the field of juvenile science fiction and his books sold well, other established science fiction writers followed suit. Notable publications from these writers included Asimov's Lucky Starr series, 1952–58, published under a pseudonym, Paul French; the Winston Science Fiction series,[18] published by the John C. Winston Company (later Holt, Rinehart, and Winston) from 1952 to 1961 and comprising thirty-five novels by authors including Lester Del Rey, Arthur C. Clarke, and Ben Bova; and the Undersea Trilogy, by Pohl Williamson and Jack Williamson, published between 1954 and 1958.[19] Many of these books featured teenage male protagonists and action taking place at academies or boot camps. Two of Heinlein's Scribner's novels were also adapted by Hollywood; *Rocket Ship Galileo* became the 1950 movie *Destination Moon*, while *Space Cadet* was the basis for the television serial *Tom Corbett: Space Cadet*.[20]

Editors working in children's literature took children's desire for science and science fiction books seriously. "Children's thinking today turns very early toward science," wrote Virginia Seaman Bechtel, founding editor of Macmillan's children's department and widely published critic of children's books, in a 1955 review of an anthology of science-themed poems. "Even beginning readers want to know about the atom, and there are books to tell them all about it. Their toys, from 'space suits' to chemical sets, reflect their mood."[21] Helene Frye, an editor at McGraw-Hill, wrote in 1964 that her favorite authors for young readers of science "recognize that the child and the creative scientist have in common not only an inquiring mind, but also a freedom from presupposition, the ability to look at things without a rigid abstract idea of what they must be like."[22]

In heralding children's advanced tastes, editors and librarians implicitly and explicitly claimed importance for their own field, which was doing such crucial work in helping children connect with science. Writing in 1959, Agnes Krarup, director of school library services for the Pittsburgh Public Schools, argued that children were ahead of the curve when it came to understanding science: "Because of rich reading resources many of our children are ahead of the average adult today in scientific knowledge, at least in some areas. Sputnik was no surprise to them; some children had been reading space travel for years." Children's librarians, because of their contact with children's books, could also count themselves as savvy predictors of scientific progress. Krarup pointed to the 1952 publication of *Across the Space Frontier*, a book with contributors that included Wernher von Braun and Willy Ley.[23] "I cite the date because I used this book among others on an adult TV show," Krarup wrote. "The man who was on the program with me was polite but skeptical, indicating privately afterward that he thought what I had said was just a bit far-fetched, more in the realm of science fiction than in the world of adult affairs."[24] After the fact, Krarup was proud to align herself with the kids, who had foreseen Sputnik.

Good science fiction, librarians argued, could recruit kids to science. In a piece titled "Literature, Science, and the Manpower Crisis," which ran in *Science* in 1957, Joseph Gallant argued that an integration of science and the humanities would create the spark of interest that would bring more students to scientific careers; about science fiction, he wrote, "Much in this medium is worthless, but the best specimens are catalysts to the imagination, not to crude and easy fantasies, but to disciplined and orderly marshaling, imaginatively, of the *possible* by extension from the known."[25] In an attempt to offer "quality" reading material, librarians tried to distinguish between this "worthless" science fiction (usually epitomized by the pulps and comics) and the "best specimens." In 1958, for example, *Junior Libraries* ran an article by author Geoff Conklin, advising librarians to look for material written by authors with some scientific training, and to "keep an open mind," not to "confuse s/f [science fiction] with cheap comics"; in fact, he added, "with a little unassuming guidance, s/f can replace comics in the reading diet."[26] New Jersey librarian Learned T. Bulman wrote in 1955 that science fiction was "bait"—introductory reading for children, shelved in the science section, that might lead young readers to more quality fiction.[27]

And Isaac Asimov, himself a scientist and science fiction writer, thought science fiction could help counter what he saw as a cultural tide of anti-intellectualism. "There is one entire branch of popular literature which is largely given over to the proposition that brains are respectable," Asimov wrote in the American Institute of Biological Sciences' bulletin in 1957. "That branch is known as science fiction." Not in movies, television, or comics, but in science fiction stories, "scientific research is presented . . . as an exciting and thrilling process; its usual ends as both good in themselves and good for mankind; its heroes as intelligent people to be admired and respected." Asimov thought that good science fiction writers who wrote for young people were, in fact, recruiters. "By way of my spare-time occupation as science fiction writer," he reported, "I now have evidence that occasionally I help to win the initial victory and encourage a youngster to go into science who might otherwise not do so. Extrapolate this to science fiction in general and think of the many youths who are won silently and who do not bother to advertise the victory."[28]

HEINLEIN'S RESPECTABLE "JUVENILES"

In 1948, a seven-year-old future science fiction writer and critic, Alexei Panshin, discovered a strange book on the shelves of the East Lansing Public Library: *Rocket Ship Galileo*, by Robert A. Heinlein. "There was nothing else in the children's collection like that," Panshin later wrote. "In the late Forties, very little serious science fiction had been published in book form for either adults or children. It was still thought of as pulp literature, more than a bit dubious in the eyes of old-fashioned small town librarians." Once Panshin began reading Heinlein's juveniles, he read all twelve of them published over the next decade, coming to see the books as guides to the real world: "It seemed there was nothing that Heinlein didn't have the true scoop on. He certainly knew more than my schoolteachers did. They were only able to teach me what was ordinary and obvious."[29]

The unusually acrimonious story behind the publication of Heinlein's juveniles, which are universally acclaimed as the most respectable and long-lasting of all the space-based science fiction written for young people during the 1940s and 1950s,[30] shows how the eager, curious space kid, the product of several different literary and scientific ideals, became a site of contested ideals of Cold War masculinity.[31] The Heinlein juveniles, which envisioned true education as a process of individualization, were profoundly influenced by the author's libertarian leanings.[32] Unlike many other authors who put space cadets at the center of lucrative book series, Heinlein emphasized his protagonists' scientific education as a central part of the narrative of his books. This education was inherently masculine. The Heinlein books were nonsequential, and each featured a different set of characters and plot requirements, but all fit into a universe Heinlein called his "Future History." The Future History is the story of humanity's emergence into a mature phase that would allow men to be able to control themselves enough to reach outward into the stars.[33] Part of this process of emergence meant the shedding of preconceived structures and notions of civilization. Learning the mental discipline necessary to be a scientist meant independence from negative, false "civilizing" forces, often personified as female unreason or moral policing.

Harking back to his own childhood as a reference point, Heinlein believed that all American boys were scientifically minded, and if they were not turning into scientists, it was because their schools were failing them. To him, the professionalization of education, its bureaucracy, and its rules were unproductive. He thought that the teaching of a true scientific mindset should include the encouragement of experimental thinking about the social order; this speculative bent often caused him to run afoul of cultural gatekeepers. For Heinlein, the "true scoop," informed by science, math, and logic, could be transformative for young people, but he believed that this transformation was obscured by the hidebound traditionalism of those who taught and wrote for children. His conflicts with his editor at Scribner's, Alice Dalgliesh, were informed by the gender politics of the postwar era; both the subject matter of his books and their publication history is the product of debates over the role of mothers and female teachers and experts in boys' lives.

Postwar worries about the fate of the young scientist in the school system often took the shape of fear that because intellectual activity was culturally perceived as feminizing—the opposite of football, and an activity that would inevitably lead a young person away from lucrative employment—no young man would want to take up science as a vocation. Science Service's depiction of successful STS finalists as athletic and physically fit was meant to counteract this perception. World War II and the postwar years were also a high point in the American discourse of "mother blaming." In these years, author Philip Wylie's much-reprinted *Generation of Vipers,* a book full of filial resentment for a culture of "mother love," was published; social scientists worried about the consequences of maternal overinvolvement for children being brought up in the "suburban matriarchy"; and Hollywood produced *Rebel without a Cause* (1955) and *Psycho* (1960), both of which showed the supposedly terrible consequences of overmothering.[34] Heinlein's project—the work of a man who identified himself as standing firmly on the side of science and scientists— was a mother-blaming, teacher-blaming postwar reclamation of science as a vital activity of American boyhood. Science was *inherently* masculine, Heinlein argued; a young scientist did not need to show his worth on the gridiron to prove that. If people perceived the situation differently, that was not the fault of science. What they were seeing was an untrue vision of science, tempered by the moralistic and idealistic boundaries imposed on it by mothers, teachers, and authority figures.

Heinlein viewed science fiction as a medium for presenting boys with a vision of a life spent committed to extrapolative thinking. He hoped readers would get hooked on the kind of "scientific" analysis that could not only provide them with power over their physical environments but also help them rethink the parameters of their social order. In twelve years of writing for Scribner's, he solidified his opinion about officially sanctioned children's literature: he resolved that those who shaped children's reading material were underestimating children's ability to handle technical detail and sociological speculation. In this, Heinlein believed, children's literature (which he viewed as the province of women, much like teaching and mothering) was failing boys; he advocated informal teaching from male elders as a way to circumvent traditional schooling. His sociological speculations within these juveniles, which included advocacy for the elimination of schools, challenged boundaries between the "adult" and the "child."

At the same time, the freedom that his young protagonists found through their independent pursuit of science was consistently gendered male. In Heinlein's vision, loving science could help a child escape the external social constraints of being young; gender, on the other hand, made inexorable biological demands, and no amount of affection for math or laboratories could save a girl from her fate.

HEINLEIN AS YOUNG-ADULT AUTHOR

In the years he wrote young-adult science fiction, Heinlein often claimed authority to know "what boys like" on the basis of his own childhood, education, and coming-of-age, leveraging his nostalgia for his own adventures as a young person in arguing for increased freedom for boys' explorations. He constructed an autobiography that highlighted his own initiative and curiosity. The author was born in 1907 in Butler, Missouri, one of nine children. The family was lower middle class, and Heinlein's father worked as a bookkeeper for International Harvester. According to his official biographer, "his father's income was never quite enough," because of his large number of children.[35] Philip Wylie, whom Heinlein admired, called himself a "motherless man," distancing his life story from the suffocating mother-son dyad he diagnosed as the cause of so many national problems; Heinlein's tales of his own youth, often full of the independence that being one of nine siblings afforded him, seemed to prove that he too made his own life.[36] Young Heinlein worked various jobs, beginning in third grade. He was an interwar science fiction fan and experimenter, reading Roy Rockwood's Great Marvel series, Tom Swift, Horatio Alger, Twain's *Huckleberry Finn*, Edgar Rice Burroughs, science fiction pulp magazines, and H. G. Wells. He bought a set of the encyclopedia *The Book of Knowledge* with money he earned delivering newspapers. He had a basement chemistry laboratory ("his youngest sister, Mary Jane, wryly remarked that her mother never knew when the house might explode") and belonged to science clubs both inside and outside of school.[37] After high school, Heinlein attended the Naval Academy at Annapolis, where he trained in engineering; he graduated in 1929 and served as an officer in the early 1930s, before receiving a discharge in 1934 for medical reasons.

Heinlein's life after the end of his active duty in the Navy fluctuated between political involvement and literary effort. Living in California, he thought he might enter politics as a Democratic candidate and worked for Upton Sinclair's End Poverty in California (EPIC) campaign. He began his writing career in 1939, selling a story to *Astounding Science-Fiction*; his writing was interrupted during World War II, when he worked again for the U.S. Navy at the Philadelphia Naval Shipyard doing aeronautical engineering. By the postwar era, he had become ardently anti-Communist; writing a sworn affidavit in 1945 on behalf of a scientist friend who needed to clear his name with an investigatory committee, Heinlein characterized himself as one who "hated" Communists and could "smell" them.[38] During the postwar years, his writing career was still nascent. Eventually, Heinlein was to garner an audience of millions for his adult and juvenile novels, which included his most famous work, *Stranger in a Strange Land* (1961); he published thirty-five books altogether, all of which are still in print today, and won the Hugo Award for best science fiction novel of the year four times.[39]

His time working with Scribner's came at a key career-building phase. Heinlein made a yearly commitment to produce a Scribner's juvenile throughout the late 1940s and the 1950s, despite his misgivings about what he saw as the restrictive nature of writing for children. The twelve Scribner's juveniles were published between the years of 1947 and 1958.[40] Heinlein also wrote a thirteenth book for Dalgliesh, *Starship Troopers*, which she found unacceptably violent and explicit and which he published with Putnam's instead. Additionally, Heinlein published *Podkayne of Mars* with Putnam's in 1963, a book with a teenage heroine that some considered a juvenile, though Hein-

lein himself did not. Many of the books were serialized before their publication — *Farmer in the Sky* was first printed in *Boys' Life*, the magazine of the Boy
Scouts. When Heinlein first started writing these books, he thought that they
would be aimed at readers in their early teenage years, with heroes three to
four years older than the readers; by the time he wrote *Starship Troopers*, he
had revised his estimate of his readers' ages upward and thought that the typical reader was "fifteen years old, male, and [had a] somewhat superior men-
tality. I am not interested in writing for dullards; I have better things to do with
my time."[41] The plots of these books were not directly related to one another,
unlike previous series fiction for young people, such as that produced by the
Stratemeyer Syndicate; Heinlein liked the idea that the books would retain
their integrity as separate novels, and he and Dalgliesh seemed to agree that
branded "series" fiction had an undesirable lowbrow flavor.[42]

Although not technically "series" fiction, the books' plots are thematically
related; all concern boys who take interplanetary journeys, arriving at new
levels of mastery and knowledge in the process. Telling the story of the writing of *Rocket Ship Galileo* for Scribner's publicity department in 1947, Heinlein
wrote that the question that prompted the writing of the book was, "In what
way will the atomic age affect most strongly the lives and interests of boys?"
"The answer," he said, "discounting the awful possibility of World War III, lay
in interplanetary flight. . . . The romance of space flight will grip the imagination of the rising generation of boys in the same fashion as did air flight for the
generation just passed."[43] Heinlein often assumed that boys would be interested in the things that fascinated him, and his interest in rocketry and space
travel might have driven this assessment of boys' interests, but in this case
the postwar fashion for space-themed television, films, radio shows, and toys
stood witness to the accuracy of his diagnosis.

The Heinlein books published by Scribner's, despite their author's roots
in the fringe world of science fiction, received critical acclaim from within
the children's literature community. In the *New York Times* in 1950, Ellen
Lewis Buell, editor of the children's section of the *Times Book Review*, wrote,
"We need more writers of the caliber of Robert Heinlein, the only author in
[the science fiction field] who writes consistently (and brilliantly) for young
people."[44] Richard S. Alm, reviewing the recent development of literature written especially for adolescents in the *School Review* in 1956, singled out Heinlein
as one of fifteen authors "making worthy contributions to the reading lives of
young people."[45] The Heinlein juveniles also appeared on the New York Public Library's *Books for Young People* list (renamed *Books for the Teen Age* in 1954)

throughout the late 1940s and the 1950s. This approbation, together with good sales, led Scribner's to value Heinlein's contributions highly; the feeling was not quite reciprocal.

Heinlein often characterized his editor at Scribner's, Alice Dalgliesh, as hopelessly conventional and timid, but Dalgliesh viewed herself as a publisher of "modern" books and a forward-thinking person. Educated at Columbia's Teacher's College, Dalgliesh initially wanted to be a kindergarten teacher before starting a career in publishing. Over her career she wrote over forty children's books, two of which (*The Bears on Hemlock Mountain*, 1953, and *Courage of Sarah Noble*, 1955) received Newbery "honor book" designations.[46] She was a longtime book reviewer for *Parents* magazine and head editor of Scribner's juvenile department.[47] During the 1920s and 1930s, when children's editors debated the relative merits of fairy tales and realistic stories (the "Grimm versus milk bottles" debate outlined in Chapter 2 of this book), Dalgliesh strongly advocated for realism. She wrote in *Publishers' Weekly*: "It is an achievement to picture everyday things with understanding and imagination. . . . Modern life . . . is full of interesting, real things, and there is no time for sugary little fairy tales of the type that used to be published by the dozen."[48] This commitment to the "new" and the "real" in this "interesting world" led her to profess great respect for science and scientists.

Female children's book editors of the 1920s and 1930s identified strongly with their profession and their community; the "bookwomen" of that era saw themselves as custodians of "taste" for the public.[49] In Dalgliesh's interactions with Heinlein, she was very aware of this gatekeeper role, often making comments and offering criticisms while protesting that she herself would not mind the more daring aspects of his juveniles, but she knew that the greater community of reviewers, librarians, and teachers would not approve.[50] The ongoing conflicts between Heinlein and Dalgliesh during his Scribner's decade highlight how the supposedly unifying project of science promotion for young people was actually full of adults with different understandings about what it meant to be interested in science—and different idealistic concepts about boyhood.

"STRONGER MEAT":
SCIENCE AND LEARNING IN HEINLEIN'S JUVENILES

In the opening pages of Heinlein's first juvenile, *Rocket Ship Galileo*, the scientist uncle of one of the protagonists inspects the group laboratory where his nephew and two friends conduct investigations. He is impressed with the level of systematization that he finds: "It is common enough in the United States for boys to build and take apart almost anything mechanical, from alarm clocks to hiked-up jalopies. It is not so common for them to understand the sort of controlled and recorded experimentation on which science is based." Uncle Don Cargreaves also visits his nephew's solitary basement laboratory, which, he tells his nephew, is a bit messy but seems to contain the makings of real science. Comparing the lab favorably to a feminized domestic space, he comments, "It didn't look like a drawing room but it did look like a working laboratory." Especially important to the scientist's positive assessment were the notebooks he saw that his nephew kept, notebooks which, one of the friends volunteers, are "the influence of Ross' old man," a retired electrical engineer. "Dad told me," Ross says, "he did not care how much I messed around as long as I kept it above the tinker-toy level. He used to make me submit notes to him on everything I tried and he would grade them on clearness and completeness." Importantly, however, Ross's dad does not continue to contribute to his son's lab activities after assessing the notebooks: "He says they [the rockets] are our babies and we'll have to nurse them," Ross tells Art's uncle.[51]

From *Rocket Ship Galileo* to his final Scribner's juvenile, *Have Space Suit — Will Travel*, Heinlein featured young protagonists who achieved manhood through hours of independent study and a devotion to exacting technical detail. His highly individualistic philosophy of education assumed that the American boy was interested in science and that what he needed to succeed at shaping his interest into commitment was not a school or a pedagogically trained educator but plenty of elbow grease and contact with older, established men who practiced science, math, or engineering. Heinlein's books preached the virtues of independent inquiry, leavened with some carefully considered contacts with "real" scientists, and promised that this approach would be rewarded.[52]

Heinlein held the libertarian belief that a school bureaucracy could not educate and that children and, to a lesser degree, their parents needed to take their education into their own hands. This point of view was informed by his own experience with schooling; he wrote that he had attended a primary school with overcrowded classrooms and insufficient facilities and that he had, nonetheless, managed to learn through his own desire to do so. Near the end of his time writing for Scribner's, Heinlein wrote to Dalgliesh defending his inclusion of corporal punishment in *Starship Troopers*; a lack of corporal punishment, for Heinlein, came to stand in for all the things he saw as wrong with the school system. Heinlein wrote: "I have formed a firm opinion that we have probably the worst secondary schools on this planet — and I have checked schools in our Deep South, the East Coast, the Middle West, and the West Coast and have compared them with schools in South America, in Singapore, in Australia, in Indonesia, in Denmark, and many other places. We have the worst schools, the most palatial school buildings, and the most over-privileged and self-pitying and under-qualified teachers I have found anywhere. And by *far* the worst discipline!"[53]

The author's objections to American schooling and American teachers focused on the gender of the schoolteacher: "Our American public schools are today largely staffed by half-educated females and spiritually-castrate males (who are just as ignorant)."[54] These teachers, he believed, were incapable of guiding young people to acquire the kind of mental discipline that was necessary for learning science and mathematics. Most effective teachers within the Heinlein juveniles are men — usually men with practical experience in the science, technology, engineering, and mathematics (STEM) fields or the military (bona fides which, by Heinlein's lights, could not help but exempt a person from being "spiritually castrate").

Part of the process of achieving mastery for the protagonists in Heinlein's books — and, thus, for his imagined readers — was the young person's realization that education was a highly individualistic project, one best pursued outside of the confining parameters set by the establishment. A young protagonist often had an epiphany in which, convinced that schooling had prepared him for a task, he would come into contact with an adult male who would convince him of the opposite. At the beginning of *Have Space Suit — Will Travel*, Kip explains that he will not receive a scholarship to MIT based on his high school education: "The emphasis [at Centerville High] is on what our principal, Mr. Hanley, calls 'preparation for life' rather [than] on trigonometry. Maybe it does prepare you for life; it certainly doesn't prepare you for CalTech." Kip finds out this hard truth when his father takes a look at his textbooks and tells him that if he continues relying on this school for his education, he is sure to flunk out if he "tackles any serious subject — engineering, or science, or pre-med" in college. Kip's dad makes an extensive chart of the various classes Kip could take in his upcoming three years in high school and despairs of his chances of emerging with a decent education. "'Son, Centerville High is a delightful place, well equipped, smoothly administered, beautifully kept. Not a "blackboard jungle," oh, no! — I think you kids love the place. You should. But this —' Dad slapped the curriculum chart angrily. 'Twaddle! Beetle tracking!

Occupational therapy for morons!'" Kip takes new classes and also starts reading books on his own ("those books were hard, not the predigested pap I got in school"). In 1957, Heinlein assessed the "themes" he had tried to "preach" in his Scribner's series; the first was "that knowledge is worthwhile in itself— and that a thorough acquaintance with mathematics is indispensable to the acquiring of much of the most worthwhile sorts of knowledge."[55] Kip, like many other Heinlein heroes, experiences the most trouble and success with mathematics. For Kip, knowledge leads to knowledge, and eventually to enlightenment, as he reads more and more math: "Analytical geometry seems pure Greek until you see what they're driving at—then, if you know algebra, it bursts on you and you race through the rest of the book. Glorious!" He begins to read about chemistry and physics and outfits his family's barn with "a chem. lab and a darkroom and an electronics bench and, for a while, a ham radio station." The end result of his self-study: he passes the College Boards his senior year.[56]

Other Heinlein juveniles also made critiques of mainstream educational systems. Part of the rite of passage in *Space Cadet*, which was based on Heinlein's own experience in the Naval Academy, lies in overcoming the insufficient prior education the boys have received in Earth schools.[57] Matt, who has joined an elite group of scientist-soldiers who keep the peace by promising to bomb any planet that threatens war, arrives at the Patrol school with an insufficient working knowledge of mathematics and needs to be brought up to speed by his tutor, Lieutenant Wong. Wong does not blame Matt but rather the system, shaking his head: "I sometimes think that modern education is deliberately designed to handicap a boy. If cadets arrived here having already been taught the sort of things the young human animal can learn, and should learn, there would be fewer casualties in the Patrol."[58] The Patrol's school achieves its educational goals through solo study, achieving, in effect, a situation much like Kip's, even within an educational institution. Most students spend time with each other only when they meet in laboratories or for group drill time. This enhances individual responsibility and accountability; as Lieutenant Wong says, "It's pleasant to sit in a class daydreaming while the teacher questions somebody else, but we haven't got time for that."[59]

Many of Scribner's juveniles written by Heinlein contained similar sequences, in which boy protagonists embark on independent study, and in these instances Heinlein is careful to juxtapose the higher value of learning done outside an official structure with the education the boys might otherwise have gotten by joining their peers. In *Starman Jones*, for example, Max

Jones studies astrogation (astral navigation) by reading his deceased uncle's books; denied entry to the astrogator's guild because his uncle failed to give him a recommendation before dying, he nonetheless succeeds at lying his way onto a ship. After a series of unfortunate events robs the ship of its formally trained astrogators, he manages to pilot the ship and its cargo of settlers back to safety.[60] Several protagonists, including the trio of boys of *Rocket Ship Galileo* and Castor and Pollux Stone of *The Rolling Stones*, turn down opportunities to study inside formal institutional structures in favor of adventure in space; though the parents, in each case, initially object, they finally agree to their children's plans when the boys agree to pursue independent study in mathematics during space flights between planets. The ensuing adventure, as well as the boys' admirable performances under pressure, conclusively answers the question of whether this choice was the right one.

Heinlein's stand against what he saw as the watering down of American education was one that he made both in the narratives of his books and in the process of writing and editing. In his engagement with children's literature, he often argued that the professionals in the field misunderstood how much technical content boys wanted and how proficient boys already were in technical subjects; Heinlein believed in the scientific authority of American boys. Writing for *Library Journal* in 1953, in an article giving tips about acquiring science fiction for young people, he implored librarians to check with "an Air Force or Artillery officer, a physics teacher, or almost any fourteen-year-old boy, especially boys who are active in high school science clubs," before choosing to buy new interplanetary science fiction for young people.[61] He argued in several different ways his point that boys could handle science fiction in which discussions about mathematics, logic, or astrogation lasted for pages at a time. It is significant that the boys' club in *Rocket Ship Galileo* was modeled on Heinlein's own high school group, the Newton Club; Heinlein often drew from his own childhood experience in claiming that boys wanted a high level of technical detail in a novel. During a conflict over the level of detail in his depiction of alien life forms in *Red Planet*, Heinlein told his long-time agent, Lurton Blassingame, who was acting as a mediator, that Dalgliesh was not qualified to assess fiction written for boys: "I've read a couple of the books she wrote for girls . . . they're dull as ditch water. Maybe girls will hold still for that sort of thing; boys *won't*."[62] He justified his position by listing his childhood activities as credentials: "Having been a boy who raised white mice, snakes, silk worms, belonged to the Scouts, science clubs, cadet corps, climbed mountains, built telescopes, radio sets, etc., I think I know a damn sight more about boy tastes than she does."[63] Failing to convince Dalgliesh that the happenings in *Red Planet* were scientifically accurate, Heinlein appealed to an outside authority, scientist R. S. Richardson, to back him up. Heinlein wrote to Blassingame, enclosing a monograph by Richardson: "It should be sufficient to stop the clock on the notion that this book *Red Planet* is an uncontrolled exercise in fantasy. Richardson's trained mind sees the implications in my details; Dalgliesh doesn't have the knowledge to see what I was doing—a direct result of the fact that she objected to overt technical explanation."[64] Heinlein implored Blassingame to try to find another publishing house for *Red Planet*, one that might be staffed by an editor "with some knowledge of science and some knowledge of the science-fiction field." "One of them," he added, "might even be a man, with recollection of what *he* liked as a boy."[65]

What Heinlein wanted was an editor who shared what he considered to be a forward-thinking allegiance to the idea of "the future"; he associated this quality with masculinity, but all men did not have it. In 1946, a (male) editor from Westminster, William Heyliger, refused a first draft of *Rocket Ship Galileo* (then titled *Young Atomic Engineers*) because it was not set in a small town but rather explored interplanetary travel. Heinlein sent a letter to Blassingame, refusing to make the revisions requested, outlining his rationale. "Boys of 1965 [when the book is set] won't be limited to a small town, unless they are either poverty-stricken or dull," Heinlein wrote. He then told Blassingame about three nephews: Buddy, "not yet out of high school, has just completed two transcontinental trips, made by motorcycle bought as junk and rebuilt by him. He wants to be a rocket pilot. He *expects* to go to the Moon." Another nephew, Colin, age 13, read some of the chapters of the book, and "his comment on the story was technical rather than literary—he said that I had not given the details of the captive test run explicitly enough. He has Boy Scout merit badges in such things as radio, electricity, astronomy, metal work, etc. Rocketry is simple from his stand point."[66] Heyliger did not see the appeal of these technical aspects, despite his gender.

This letter set a precedent for Heinlein's ongoing crusade to convince Dalgliesh that kids wanted more science; he gathered anecdotal examples from all over. Heinlein asked a friend in New York to monitor a radio show called *Young Book Reviewers* for a review of "Space Cadet," and the friend replied:

> I called WMCA and they tell me the program reviewing your book
> was one of the best they had. . . . The Bronx Science High School
> loosed a gang of science fiction fiends on the program and they

had a hell of a time. Even the woman who runs the program was impressed—she allowed that she might have to look into science-fiction sometime. I might add that the station does not select the book. It is picked by the youngsters and yours was the first of its kind to be reviewed. Prior to that there had been derogatory comment about "Buck Rogers stuff" but the s-f boys persisted and they tossed them a bone which turned out to have a lot of meat on it. I rather gather that more s-f will be used in the future.[67]

Heinlein argued that part of the reason why older people did not understand science fiction was that they were out of touch with advances in science and incurious about its workings. Children, on the other hand, such as his characters in the Scribner's series, are "commonplace 'heroes,' surrounded by the technology of their period—but they aren't impressed by it, at least no more than a kid today would be impressed by a trip to the Boulder Dam. They take the gadgetry for granted, as I took the telephone for granted, and as the current crop of youngsters take TV for granted."[68] In a 1951 letter to his aunt Anna Lyle, a retired schoolteacher, Heinlein promised to send along a copy of *Space Cadet*, saying, "I think it is one you would like even though it is on my boys' list—or perhaps because it is. I feed the boys stronger meat than I do adults. Most adults just don't have as much intellectual curiosity as the kids do and I have to bear that in mind when I write for adults. With the kids, when I get an interesting idea, I can play with it."[69]

SOCIOLOGICAL SPECULATION AND
THE DISRUPTION OF AUTHORITY

If Dalgliesh objected to the technical detail present in some of Heinlein's writing for boys, she had even more trouble accepting the sociological speculation that later became his trademark contribution to science fiction. This conflict caused more trouble between the writer and his editor, as Heinlein believed that the social sciences were even more interesting—and worthy—as a topic of inquiry than were the natural and physical ones. Explaining the Academy's processes of education, *Space Cadet*'s Lieutenant Wong tells the hero, Matt, about "hypno," a procedure that will help him study certain subjects under the influence of hypnosis. "Everything that can possibly be studied under hypno you will have to learn that way in order to leave time for the really important subjects," he says, to which Matt replies, "I see. Like astrogation." Wong then counters, saying that astrogation is "kindergarten stuff," telling Matt that he knows "from your tests that you can soak up the math and physical sciences and technologies." What's more important to learn with a conscious mind, Wong says, is "the world around you, the planets and their inhabitants — extraterrestrial biology, history, cultures, psychology, law and institutions, treaties and conventions, planetary ecologies, system ecology, interplanetary economics, applications of extraterritorialism, comparative religious customs, law of space, to mention a few."[70]

Many of the tests of mental agility that Heinlein's protagonists undergo are not strictly "scientific" but rather involve larger questions about culture and social arrangements. Heinlein believed that "speculative fiction," a term he preferred to "science fiction," "is also concerned with sociology, psychology, esoteric aspects of biology, impact of terrestrial culture on the others [*sic*] cultures we may encounter when we conquer space, etc., without end."[71] The boys of *Space Cadet* were supposed to use the habits of mind they had acquired from studying "hard" science to face the more difficult questions of the "soft" sciences.

Many of the extrapolative situations that Heinlein's juveniles favored played with the arrangement of social relations, in order to provoke his young readers into asking questions about the fungibility of human nature and the ramifications of cultural change; it was these that caused the greatest impasse between him and Dalgliesh. During the conflict over *Red Planet*, Heinlein wrote to Blassingame that he thought Dalgliesh objected to the book because she was fundamentally confused about the nature of science fiction. Because Dalgliesh was proud of having published Kenneth Grahame's *Wind in the Willows*, Heinlein said, she did not object to fantasy or fairy tales; rather, "she has fixed firmly in her mind a conception of what a 'science-fiction' book should be, though she can't define it and the notion is nebulous." Heinlein thought Dalgliesh's ideas about science fiction could be summarized thus: "Science has to do with machines and machinery and laboratories. Science-fiction consists of stories about the wonderful machines of the future which will go striding around the universe, as in Jules Verne."[72] Although his fiction did contain some laboratories and wonderful machines, he thought this approach was uninteresting; fiction that described machinery without including speculation about the changes in human relations in the society surrounding that machinery was less fully realized than his own efforts.

Unfortunately, he found, imagining possible future social orders often put him in conflict with Dalgliesh and others within the field of children's literature. The fundamental conflict, he believed, was between the mission of speculative fiction, which he articulated as the exploration of the assumption, "The customs of our tribe are *not* the laws of nature," with editing suggestions that, he argued, "add up to the notion that I must never assume that the present-day customs, opinions, and attitudes dear to our tribal shamans are anything less than divinely inspired and immutable."[73] Although Dalgliesh wrote in *Publishers' Weekly* in 1943, in a wartime essay about children's books and the teaching of "tolerance," that children's literature needed to incorporate "fully thought out factual material, history, biography, and — much needed — anthropology" in an effort to teach children about "differences and prejudices," she believed in the explanation of present-day "truths" about people across the globe.[74] In contrast to this latitudinal approach, which assumed that tolerance could be taught by showing children in one stable, unified culture the truth of how others living in faraway stable, unified cultures existed, Heinlein's extrapolative approach, which merged science fiction with anthropological and sociological speculation, was meant to make children think about the long view and changes in social customs over time. Because of the particular themes that Heinlein chose for his extrapolations, which included age, family, race, and gender, this aspect of his Scribner's juveniles often caused friction with his editor.

Some of the conflicts that Heinlein anticipated, but which barely arose, were over his message regarding tolerance of other races and cultures. He officially espoused a stance popular in midcentury liberalism: color-blind inclusionism. In this, he was part of a trend in children's books of the 1940s (as Dalgliesh pointed out in her essay of 1943, this was a major theme of wartime book production; as Mary K. Eakin wrote in 1955, in the years after 1945 "scarcely a [children's] fiction book was published that did not either treat an aspect of intercultural relations as its main theme or else bring in some problem of this sort as a subplot").[75] In a letter to Blassingame written while he was working on *Young Atomic Engineers*, Heinlein wrote that his heroes were of Scotch English, German, and American Jewish extraction and warned, "You may run into an editor who does not want one of the young heroes to be Jewish. I will not do business with such a firm. The ancestry of these three boys is a 'must' and the book is offered under those conditions. My interest was aroused in this book by the opportunity to show to kids what I conceive to be Americanism."[76] The conflict did not arise, perhaps in part because Morrie, of *Rocket Ship Galileo*, was never explicitly identified as Jewish, despite the presence of certain vaguely invoked cultural and religious markers.[77] During the editorial process for *Tunnel in the Sky*, which included a prominent black female character, Caroline, Heinlein told Dalgliesh that he "wanted Caroline identified as Negro from the start. . . . This girl's characterization all through the book is believable only if she is colored, I want her tagged from the start." Replying to Dalgliesh's concern that "this Negro secondary character would lose us sales in the South," he wrote back, "This is not a point on which I am willing to budge."[78] He did, however, change the identifier used to describe Caroline from "black" to "Zulu," thereby giving her an exotic provenance that would explain her "characterization" as a brash, uncouth female warrior while also abstracting her from present-day conflicts in the United States. While Heinlein defended Caroline's right to exist, he marked her as different: loud, violent, and romantically unfit (important in a book whose plot included more than a few romantic pairings).

Heinlein's extrapolations regarding the relative power of parents over children and of children in society were especially troublesome for Dalgliesh. One of her major issues with *Red Planet* was a scene in which the boy protagonists use guns to fight alongside their fathers in a colonial uprising. In a meeting in which the adult members of the colony resist the idea of arming what he considers to be "children," another colonist argues: "This is a frontier society and any man old enough to fight is a man and must be treated as such—and any girl old enough to cook and tend babies is adult, too." Referring to their coming conflict with the corporation that owns their land, he adds, voicing Heinlein's true beliefs: "Whether you folks know it or not, you are headed into a period when you'll have to fight for your rights. The youngsters will do the fighting; it behooves you to treat them accordingly. Twenty-five may be the right age for citizenship in a moribund, age-ridden society back on Earth, but we aren't bound to follow customs that aren't appropriate to our needs here."[79] Heinlein, himself a firearms buff, wrote to Dalgliesh that he believed in "training kids early in the use of 'dangerous' weapons."[80] The issue of armed heroes arose again and again; in 1953, Dalgliesh asked him to remove from *Farmer in the Sky* a reference to the protagonist defending himself against his stepfather's beatings with a knife—a request that provoked a three-page disquisition from Heinlein on the virtues of an armed citizenry. In 1958 she objected to a scene in *Have Space Suit—Will Travel* in which the hero Kip throws a malted milk in a bully's face.[81] Heinlein asked Dalgliesh to let him know whether she actually received objections to this scene from librarians and wrote (dramatically): "If the control over what teen-agers read . . . actually is in the hands of persons so tender-minded, so pacifist, so conformist, so emotionally and spiritually castrate as this would imply, then I must seriously reconsider what use I want to make of the remainder of my life—whether to retreat no farther but fight them on their own ground, or whether to seek other battle ground of my choosing."[82] Heinlein blamed "weak-stomached ladies of both sexes, tender-minded creatures who fear fighting more than they fear slavery," for this censorship: "They have done their damndest to raise up a generation of sissies, afraid to fight, not trained to fight."[83] These conflicts may have been part of Dalgliesh's ongoing effort to keep the books above the "comic book level"; violence was a genre trademark she wanted to avoid. Heinlein viewed the restriction as a serious imposition—a perceived feminization he resented so much that he threatened to stop writing for young people altogether.

Heinlein's protagonists armed themselves not just with guns or malted milks, but also with the law, and this particular brand of speculation also caused the author trouble. In *The Star Beast*, published in 1954, right at the time the Kefauver Committee debated comics regulation, the protagonist John Stuart Thomas divorces his mother, who wants him to become a lawyer rather than journey to a faraway planet and become an expert in xenobiology. The character of Thomas's mother violates all of the moral goods that Heinlein has designated: she clings to tradition and so demands that her son get an accredited law degree instead of learning in an informal environment; she refuses even to try to become interested in xenobiology; she attempts to leverage guilt about leaving her side to keep him from growing up. In her self-centered superficiality, Mrs. Thomas embodies many of the qualities of Wylie's toxic "momism." The portrayal makes for a strong contrast when compared with the family-oriented vision of other science-fictional narratives of space travel that were popular in the postwar period.[84]

Learned Bulman, a children's librarian for the Free Public Library of East Orange, New Jersey, wrote to Dalgliesh to inform her that he planned to give the book a negative review in the *Library Journal*, based on this plot point ("It is certainly one of his best but WHY did he destroy it with his reference to the Court of Divorce for Children?").[85] In the ensuing flurry of letters, Heinlein told Bulman that he believed that society's inevitable progress toward a "better civilization" would certainly make it common for children to have the right to initiate a split with parents: "I expect the idea of minor children as persons in the eyes of the law to grow in the same fashion in which we have seen women become legal persons."[86] This episode added to Heinlein's dissatisfaction with the process of writing juveniles—he felt that Dalgliesh had not defended him sufficiently and wondered if he should stop writing for young people altogether if he was going to be "stuck in a dead-end street with a barricade marked 'Librarians' across it"; two years later he was still considering whether a plot change would bring "librarians such as that pompous and provincial Mr. Bulman baying at my heels."[87] The conflict also exemplifies Heinlein's belief that young people's (boys') opinions should hold more weight in the adult world. "Do you really think that children are morally obligated to be humble to adults even when the adults are utter fools?" he asked Dalgliesh.[88] Bulman, for his part, stood his ground, writing in a generally positive appraisal of the young-adult science fiction genre in *Library Journal* in 1955 that his library turned down *Star Beast* because the Court of Divorce for Children was "tossed in with very little explanation and, to top it off, a fresh brat is presented as an example of its usefulness."[89]

If Heinlein succeeded in imagining multicultural friendships, armed young people, and child-parent divorce courts, the results of his depiction of gender relations are more complicated and point to a belief in biologically determined gender roles. Much has been written about Heinlein's female characters, and critics seem to concur that representations of girls and women in his juvenile fiction are a mixed bag. Science fiction author and critic (and 1953 STS finalist) Joanna Russ, writing in 1972, argued that Heinlein's work fails at imagining alternative futures when it came to gender roles, while critic Marietta Frank maintains that the intellect of female characters in *Have Space Suit—Will Travel* and *The Star Beast*, along with the alien matrilineal societies of *Space Cadet* and *Citizen of the Galaxy*, render Heinlein's work quasi-feminist.[90] Heinlein argued that girls could and did read and enjoy his books; he dedicated *Red Planet* to his niece, Tish, who, he wrote to Dalgliesh's assistant Virginia Fowler, "is one of my most loyal fans . . . like many little girls, she reads boys' books as much or more than girls' books. (This seems to be a fact and should be significant to publishers; Ginny [his wife] tells me that she borrowed and read all the Tom Swift books et cetera because she found girls' books unbearably dull.)"[91] In a 1951 letter to Dalgliesh, he mentions reading fan letters and says, "I think it might interest you to know that I get more letters from girls about my 'boys' books than from boys."[92]

Heinlein's attitude toward female training in science, technology, engineering, and mathematics was, on the surface, a positive one. During World War II, Heinlein scouted universities for female engineers, looking for draft-exempt workers for his research division at the Philadelphia Naval Yard; he wrote that he found at the University of Delaware that the School of Engineering did not permit females to register and was furious: "I took nasty pleasure in chewing out the President of the University . . . by telling him that his University's medieval policies had deprived the country of trained engineers at a time when the very life of his country depended on such people."[93] Virginia Gerstenfeld, whom Heinlein met during the war when they were both employed at the Naval Yard, was a chemical engineer, and when they married, Heinlein wrote to a friend that she "knows more science than I do," following up by praising her for also possessing more traditional female traits: "she is an excellent cook, a good housekeeper, and a good money manager."[94] After they were married, Virginia stopped working as a chemist. Like the female STS finalists who dropped into a different career track when they married, she was a woman who heartily appreciated science but did not actively practice it.

These young women, so interested in science or soldiering or engineering, illustrate the contradictions within Heinlein's libertarian philosophy of thought: while the investigation of scientific subjects is endless and the boy protagonists discover only more frontiers to pursue, the story of a girl growing up often hinges on her decision to forsake further study—to accept limitation, rather than expansion. Virginia Heinlein seems to be the ideal real-life counterpart to (or model for?) Heinlein's fictional girls, who often exhibit real competence at various "manly" jobs, only to forsake these occupations once they "grow up" or get married. Betty Sorensen, John Thomas Stuart's girlfriend in *The Star Beast*, acts as a lawyer throughout the book, managing her boyfriend's efforts to keep his xenopet Lummox from being destroyed by fearful townspeople; at the end of the book, however, to the amazement of some adult onlookers, she sublimates her pushiness and savvy to John Thomas's future career, bargaining with authorities until he is promised a post as an ambassador. In *Tunnel in the Sky*, Rod, the protagonist, has a sister who is an efficient soldier but who gives up her career at the end of the narrative to get married.

Perhaps the best example of Heinlein's vision of scientific girlhood is his book published by Putnam's in 1963, *Podkayne of Mars*. In the only one of Heinlein's juvenile books to feature a youthful female protagonist, the spitfire Podkayne (Poddy) Fries, Martian colonist and daughter of a professor father and engineer mother, who has her heart set on being the captain of a spaceship. Her adventures accompanying her brother and great-uncle on a tour of Venus and Terra, however, end in trouble, as she is kidnapped and injured by a fellow passenger. Poddy meets a young man who also wants to be the captain of a ship and slowly starts to think that she should "let a man boss the job, and then boss the man," instead of striving for the job herself. Looking at herself in the mirror, checking out her newly broadening figure, she muses, "One might say we were designed for having babies. And that doesn't seem too bad an idea, now does it?" Remembering an emergency on board the ship when she helped out in the nursery, she thinks, "A baby is lots more fun than differential equations." She reflects that, instead of training to be a pilot, she might train to engineer the pediatric departments (crèches) of starships, thinking, pragmatically, "Which is better? To study crèche engineering and pediatrics — and be a department head on a starship? Or buck for pilot training and make it . . . and wind up as a female pilot nobody wants to hire?"[95]

Originally, the story ended with a scene in which Poddy dies trying to protect a juvenile native; the protest from the editor of *Worlds of If*, where the story was first serialized, led Heinlein to rewrite the ending so that it was somewhat more ambiguous before publishing the tale in novel form.[96] Poddy's tragic end, one that no male Heinlein protagonist ever had to suffer, epitomizes the closing off of her trajectory toward self-realization — a closure that no degree of intellectual interest in science could prevent.

CONCLUSION

Heinlein's final book for Scribner's, *Starship Troopers*, is the locus of his most public and notorious conflict with Dalgliesh. In this juvenile, Johnny Rico, a wealthy young man, joins the military after graduating from high school. Earth is unified under one government and has one military, and the franchise and political office are restricted to veterans, who are presumed to have a superior understanding of what being a citizen really means. The narrative follows Rico through boot camp, where he learns lessons about the value of submitting to properly constituted authorities, and into war against a communistic interstellar enemy, the Bugs.[97]

At first glance, *Starship Troopers* seems to be a digression from the rest of the Heinlein juveniles, in its focus on war rather than science. Rico does attend officer candidate school, where he struggles to acquire enough mathematical skills to pass his tests, given his earlier poor preparation in high school. However, the substance of the book is not about Rico's scientific education but about his transformation from callow youth to military man. The authority figures in this story do not encourage Rico to keep more extensive lab records; they teach him instead that a failure to follow orders to the letter will earn him corporal punishment. Heinlein, like many adults of his era, was worried that juvenile delinquency threatened the nation's future; however, unlike some critics, he did not believe that the mass media had caused the problem, asserting that the lack of corporal punishment in schools and at home was the cause.[98] Dalgliesh wrote to him: "Do you *really* believe that flogging is the remedy for juvenile delinquency?" She reminded him: "Some of the boys have been beaten all their lives. What about the social conditions under which these boys have to grow up in cities—living often as a family of eight or ten in a room?"[99] Heinlein wrote back: "I meant what I said in that book, namely that the almost total abolition of corporal punishment in schools and the great fall-off in same in the home, all the direct result of the pernicious influence of a school of self-styled 'experts,' is a major factor in the present decay of our Republic."[100] Making a distinction between random beatings and corporal punishment, "applied judiciously according to an explicit code of conduct," he compared the two to "rape and marital love."[101] The prevailing drift toward permissive parenting in the second half of the twentieth century has encountered significant resistance in the form of conservative backlash.[102] While the scientist-parents and parents of STS winners that Marianne Besser interviewed advanced this ideology of companionate home life as a cornerstone of scientific inquiry, Heinlein believed that intellectual and physical discipline went hand in hand.

Heinlein's views on militarism had shifted over the two decades in which he wrote Scribner's books. When he first published *Rocket Ship Galileo*, he wrote in a biographical summary for the Young Literary Guild's magazine *Young Wings* that space travel might bring "a new feeling of global unity. When men begin to think of Earth as their home, as they now think of America, or India . . . there might be a curious and very wonderful psychological result—it is possible that this opening of the Age of Interplanetary Exploration might be the end of the Age of War."[103] But by the time *Starship Troopers* came out, Heinlein had become a more militant Cold Warrior; one newspaper interviewing him on the occasion of Sputnik reported that Heinlein "commented acidly" on the lack of American foresight: "'It was the same thing with the atom bomb and the H-bomb,' Heinlein said bitterly. 'We have been living in a fool's paradise. When will we believe them when they say they are going to do something? The announced aim of the Communist party and the Soviet Union has been to take over the world.' He made a sour face. 'Look at the map today.'"[104] To another reporter asking about Sputnik, Heinlein said, "This isn't a matter of prestige—it's a matter of saving our necks. It's not even a question of can they clobber us, but how much intercontinental hardware is ready over there? Will they hit us this afternoon? Next spring?"[105]

Starship Troopers was a post-Sputnik book, intended to prepare young men to fight. The militarism of the book can be seen in the titles Heinlein suggested to William McMorris, who edited the book for Putnam's: "Shoulder the Sky" (from A. E. Housman: "The troubles of our proud and angry dust / Are from eternity, and shall not fail. / Bear them we can, and if we can we must. / Shoulder the sky, my lad, and drink your ale."); "Better to Die" (Horace Gregory: "Better to die / Than to sit watching the world die"); and "Dulce et Decorum" (the classic from Horace: "Dulce et decorum est pro patria mori," or "How sweet and fitting it is to die for one's country").[106] In a much-revised letter to Dalgliesh, just before he left Scribner's, Heinlein wrote: "What I am saying to my young reader is: 'Look, son, this is not an easy world; this is a grim and dangerous world—and it is quite likely to kill you. But you have a free choice: you can go to your death fat, dumb, and happy and never understanding what is happening to you right up to the time the bombs fall ... or you can grow up, face up to your harsh responsibilities, look death in the face and defy it, and thereby enjoy the austere but very real and deeply satisfying rewards of being a man. But the choice is yours, and neither your mother, nor your teacher, nor the state can in anywise relieve you of it."[107] This is the connection between *Starship Troopers* and the earlier juveniles: in all these books, Heinlein visualized himself as reaching out to boys, circumventing the authority of the mothers, teachers, and the state, to provide them with what was "true"—whether that "truth" lay in a commitment to military service or a commitment to science.

In a 1940 story called "Requiem," published in *Astounding Stories*, Heinlein's character Delos D. Harriman, an ancient capitalist, buys a trip to the moon, where he intends to die. Recalling his youth, Harriman muses: "There were lots of boys like me—radio hams, they were, and telescope builders, and airplane amateurs. We had science clubs, and basement laboratories, and science-fiction leagues—the kind of boys that thought there was more romance in one issue of the *Electrical Experimenter* than in all the books Dumas ever wrote. We didn't want to be one of Horatio Alger's get-rich heroes either; we wanted to build space ships." The dreaming hero then imagines his nagging wife calling him in from space, as he floats toward the moon in his rocket: "Delos! Come in from there! You'll catch your death of cold in that night air."[108]

Just as the organizers of the Science Talent Search visualized young "creative scientists" languishing in an inhospitable peer culture, Heinlein thought that curiosity was being curtailed by the weight of convention. This female intervention into Delos's reverie symbolizes the forces of conventionality that, he believed, keep boys — and men — away from true learning and advancement. In the Cold War era, the nostalgic older scientists and engineers who thought of their own childhoods as utopian spaces of learning and inquiry sought to re-create those spaces for the rising generation. Heinlein thought of himself as a true advocate of the scientifically minded American boy, believing that shared qualities of gender and scientific interest transcended his editor's professed expertise in children's culture. His books — the intellectual heart of a space-mad children's popular culture — reinforced these boundaries with vigor.

5

The Exploratorium
and the Persistence
of Innocent Science

In 1977, Amy Carter, presidential daughter and self-described "science freak," visited the Exploratorium in San Francisco, while her mother attended to official business in the city.[1] Newspapers reporting on the freckled ten-year-old's visit to the museum described it as a happy whirl. "For nearly three hours Amy made music on a dance floor, created an 'avalanche' with ball bearings, watched her face through a glass as it turned into the face of a friend, yelled 'Yeeeaaaa' into an echo tube and saw people shrink in the 'distorted room,'" wrote a reporter for the *San Francisco Examiner*. The contrast between Amy's childlike enjoyment of the space and the official trappings of her visit was irresistible: "As she skipped, limped, walked and fidgeted around the Exploratorium, with a cloth daisy sewn on the bottom of her new blue jeans, three straight-faced Secret Service agents followed, occasionally toying with the air pump and a logarithm exhibit." Photos of Amy holding hands with a local seven-year-old who had been drafted as a tour guide made sure to feature that cloth daisy. Amy Carter, a socially aware child of the 1970s who was famous for her interest in books and animals, was in her natural element in San Francisco's experimental Exploratorium.[2]

The Exploratorium, founded in 1969 in the city's Palace of Fine Arts, was purposefully anarchic, utopian, and carnivalesque in its design, encouraging visitors to touch, to manipulate, and to follow their own paths through what it called a "forest" of exhibits. In 1973, one magazine writer described a few of these: "A rope strung 120 feet across the building with an attached cord hanging down. Pulling it causes giant waves to race along the rope.... A suspended, rotating chair and a large gyroscope for the seated person to hold. ... A treelike sculpture of thousands of lights whose brilliance is dependent on how much loud noise it hears (from you)."[3] The whole place encouraged loudness, which it viewed as a sign of engagement; one reporter described the result, noting "shrieks of glee [ringing out] from the small scientists" roaming the "forest" floor.[4]

"The Exploratorium is really a scientific funhouse, a giant experimental laboratory," wrote a reporter for the *Los Angeles Herald Examiner* in a 1979 article recommending the place to tired parents touring the Bay Area with families. Stretching for explanations, the reporter wrote, "It's the equivalent of a mad scientist's penny arcade."[5] Others trying to sum up the Exploratorium in the 1970s and 1980s called in evocative descriptors that drew on cultural clichés about other places where joyful science might happen: "A science fair gone blissfully mad," "an eccentric physicist's attic," "a scientific Disneyland."[6] The founder of the place, Frank Oppenheimer, himself bearing many of the characteristics of the gleefully unhinged experimenter, often said: "Discovering the Exploratorium is like stumbling into the belly of a giant whale where some mad scientist has found a home."[7]

Although the Exploratorium is often credited with inventing the idea of the "hands-on" exhibit, the museum sought more than simple interactivity, looking to become a well-engineered safe space where people could lose their fears and doubts while learning. Oppenheimer often wrote that he wanted an exhibit to be cheap enough that it could be assembled quickly and handled roughly without fear of loss. This characteristic would remove staff members' fear that young people would do harm to the physical plant and eliminate subtle cues that kept young visitors to typical museums from losing themselves in exploration. The museum *employed* young people, as well as receiving them as guests; in the museum's understanding of its own ecology, the adult staff should step back and let natural fascinations take their course. Teenage interpreters — bright and engaging people who were not experts in science but were willing to learn along with the visitors — were the perfect solution, offering assistance and inspiration without being too threatening.[8] Other staff members came with a wide array of qualifications; like Oppenheimer, many were not museum professionals, and their choices to become involved with the Exploratorium were more a matter of affinity than a conventional desire for professional advancement. This motley makeup of physical plant and staff added to the place's romantic image. "[The Exploratorium] is not run by industry or government," wrote a reporter in 1982, "but by an improbable assortment of high-school students, scientists, educators, artists, and dreamers."[9]

The Exploratorium—which constantly reiterated its desire to welcome people of all ages—was bent not only on science promotion among kids but also on holistic social restoration. As public unease over the ecological crisis and the Cold War nuclear arms race grew, Oppenheimer, who wrote and spoke often on the relationship between science and social issues, diagnosed a problematic detachment among his fellow citizens: "I've felt that most people were getting used to the idea that they couldn't understand things, and that's very bad, whether it's about politics or nature."[10] Drawing from the language of humanistic psychology, Oppenheimer argued that the Exploratorium was a place where science, lately a threatening force operating far from most people's daily lives, could be reunderstood as personal, restorative, and communal, tapping into each person's capacity for curiosity and learning.[11] The place's association with a manic, rainbow vision of free childhood play contributed mightily to that image.

TARNISHED SCIENCE IN CHILDREN'S CULTURE

The Exploratorium was founded at the beginning of a decade that would mark a low point in adult American feelings toward science and technology.[12] Countercultural critiques of perceived scientific hubris gathered steam during the 1960s, fueled by antiwar sentiments and a growing awareness of ecological crisis. By 1970, the same year after the first man landed on the moon, enthusiasm for the space race—a driver of popular scientific enthusiasm since the postwar years—had significantly diminished. Americans were increasingly uncompelled by big symbolic acts of collective technological effort and interested instead in earthly matters of social justice and personal subjectivity.[13] Dire predictions related to pollution and population growth preoccupied writers of scientific nonfiction and science fiction; ecological crisis served as plot backdrop for several terrifying movies; and activists spoke in apocalyptic terms about the possible effects of continued overdevelopment and technological hubris.[14]

The youth movement of the 1960s mounted critiques of science as part of larger arguments against the transgressions of capitalism and war. In Students for a Democratic Society's 1962 Port Huron Statement, the use of advanced physical science for war, rather than social good, was termed a disappointing "paradox": "With nuclear energy whole cities can be easily powered, yet the dominant nation-states seem more likely to unleash destruction greater than that incurred in all wars of human history."[15] The activists situated the technological achievements of their culture firmly inside that human history, refusing to admire scientists' accomplishments simply on their own merits. Paul Goodman wrote in his 1960 polemic *Growing Up Absurd: Problems of Youth in the Organized System*, which was to become a much-referenced book for activists in the youth movement of the 1960s, that the secrecy and competition of the Cold War had similarly "corrupted" the adventure of space exploration. He thought that the younger generation was particularly harmed by this loss and asked incredulously: "Our government cannot see that noble things must not be made base, romance must not be turned into disillusion, or what will become of the young people?" Goodman was disturbed by science's postwar status as a proper career for those aspiring to an accepted model of middle-of-the-road success. Pointing to the demographic profile of recently selected astronauts ("all prove to be white Protestant, in their early or middle thirties, married, with small children, and coming from small towns—in brief, models of salesmen or junior executives for International Business Machines"), Goodman thought scientific excitement had been co-opted and scientific acumen had been remade into a standard-issue tool for achieving social conformity.[16]

While radical college-age youth rethought their image of science and technology, children's popular science changed too. In 1974, McGraw-Hill science editor Howard E. Smith Jr. asked about younger students' attitudes toward science, summed up his perspective for an interviewer for *Library Trends* magazine:

> Young people look around them and see pollution. They often blame science and technology for the pollution they see. In the 1920s, and up to 1950 or so, science readers read books on how to make a shortwave radio set and they would say, "Wow! Isn't that great? Isn't science marvelous? Think of all the things science will bring us!" Today the attitude is very different. They know what science is bringing. It has brought, in some kids' minds, pollution. They feel the Viet Nam War was run and operated by science, technology, and certain economic factions in the United States. Many of them feel that we have put a lot of money into the space program and not into urban problems.

> Maybe men did walk on the moon, but young people have to walk among garbage cans and junkies, and they see the difference.[17]

It is almost impossible to separate *adult* preoccupations about nuclear war and the environment from their diagnoses of children's worries. In the 1970s and early 1980s, activist-minded adults such as Smith were convinced that children shared their concerns about ecological collapse and nuclear war and used stories of children's fears as spurs to action. Jimmy Carter famously cited his daughter Amy's opinion in a debate during the 1980 election, mentioning that she had marked "the control of nuclear arms" as her most important issue when he asked her for her input—a remark that left the president open to accusations of softness for taking direction from a child.[18] By 1983, when ABC ran the two-hour made-for-TV film "The Day After," antinuclear activists publicly worried that children would be permanently scarred by the fear of nuclear war and used young people's adverse emotional reactions to the arms race and the specter of nuclear annihilation as political talking points.[19]

Anecdotally speaking, we do have evidence that 1970s children were, indeed, scared of ecological collapse and nuclear war. In her 1998 memoir, writer Joyce Maynard recounted the story of reading Paul Ehrlich's book *The Population Bomb* as a high school student in 1969. She felt, she wrote, "not personal, individual fear, but end-of-the-world fear, that by the time we were our parents' age we would be sardine-packed and tethered to our gas masks in a skyless cloud of smog."[20] In a letter to Senator Gaylord Nelson (organizer of the first Earth Day) in 1970, student Tom Kubald echoed this existential terror: "I think by the year 1975 every person will be Dead. More jets, mills, and cars will be built. It will make Wisconsin more polluted and the whole world will be polluted. Right now people are dying of pollution. So would you please take this into consideration."[21] Thirteen-year-old DeDe DeMuth of Baltimore, Maryland, expressed a sense of powerlessness and rage when she wrote to the Earth Day organizing committee: "What can a 13-year-old do about air pollution and what power do I have over big men who smoke cigars, probably nothing. What can I do about exhaust from cars and trucks. . . . It is my generation who will have to live in the smog and I have even heard that Florida will be uninhabitable in 20 years."[22] The Earth Day festivities of 1970 featured many high school and elementary school participants like Kubald and DeMuth. These students formed advocacy organizations with silly names, such as SLOP: Student League Opposing Pollution (State College,

Pennsylvania) and YUK: Youth Uncovering Krud (Schenectady, New York). Some names also reflected the fear and guilt that young people felt about environmental crisis: SCARE: Students Concerned about a Ravaged Environment (Cloquet, Minnesota); SHAME: Studying and Halting the Assault on Man and the Environment (Richmond, Virginia).[23]

Children's books and media published during the 1970s echoed these anxieties over the direction of technology and science. Twentieth-century children's culture, up to this point, had taken the bigger adult worlds of industry and technology to be infinitely fruitful classrooms: sites where children could learn about scientific thinking, develop a sense of the order of things, and become generally intrigued and stimulated by the world. Postwar children's literature continued the general interwar tone of positivity about human development and infrastructure, with volumes featuring friendly bulldozers and snow-shovel protagonists; many of the authors of these books received assistance from representatives of business and industry in their research.[24] By the 1970s, this pro-development, pro-technology mood had changed. While the simple story of Dr. Seuss's *Lorax* (1971), in which the Once-ler's greed threatens the livelihood of the truffula trees, is the most commonly remembered environmental tale from the early 1970s, the years around the first Earth Day yielded many other picture books and magazines that told dark stories about pollution, consumption, and waste. This media inverted a set of values previously used in promoting scientific activity, some even questioning the ultimate usefulness of creativity and curiosity in encountering the world. (In Bill Peet's *The Wump World*, a 1970 book that featured a race of spacefaring Pollutians who invade and take over the gentle Wumps' pastoral planet, the wherewithal the Pollutians show in figuring out interplanetary travel matters much less than their greed in annexing its resources for their own.) In a dramatic about-face that reflects a new adult awareness of the "age of limits," these books ask child readers to think about unmaking, rather than making.[25]

At its darkest, children's literature of the 1970s asked children to imagine the end of the natural world. "This is a scary story," reads the introduction to science fiction author George Zebrowski's "The Firebird," published in the October 1979 issue of Ranger Rick. "It tells about something that *might* happen far in the future. But it *won't* happen if we do everything we can to protect our world and wildlife. Imagine it is the year 2000." The reader hears of a girl named Melanie, who dreams about seeing "firebirds," her name for flamingos. Pathetically, Melanie knows about the birds only because of "pictures she had seen in an old book." Zebrowski's narrative borrows from several tropes of adult science fiction of the 1970s: Melanie needs to wear a smog mask and eye protector, and it is hard to see in the city "because of the thick, poisonous smog in the air." She visits a national park, with the desperate hope of seeing the firebirds, but leaves without success; the firebirds, says a ranger, are definitively extinct. The story closes with a note from the ever-present Ranger Rick, the editorial voice of the magazine, which brings the reader into sympathy with Melanie and her world: "How old will you be in the year 2000? What do you want your world to be like then?"[26]

Though this grim short story was singular in its form (very few other science fiction stories were published in *Ranger Rick* during this decade), its approach to teaching children about extinction and pollution illuminates much about the magazine's stance toward these dark issues during the 1970s. The magazine enlisted children as animals' sympathetic allies, arguing through its long-running fictional serial, "Ranger Rick and His Friends," that children and animals wanted the same things (fresh air, wild spaces, and room to range) — all of which were under threat.[27] The magazine depicted the threats to animals and children — technology, capitalism, and the greed of "grown-ups" — as monstrous and uncontrolled, representing the most irrational tendencies of the human race. If animals' problem is "people," as many of the magazine's pieces argued, how could young readers, after feeling such sympathy with animals, reconcile themselves to their membership in the human race?

Dark stories like these appear all the more striking when compared with the optimism of Robert Heinlein's juvenile science fiction books, published two decades before, or the cheerful technological boosterism of children's interwar nonfiction. If science was not always explicitly named as the "bad guy" in these stories, some older ways of "making kids love science" — discussion of space travel; demonstrations of "chemical magic" — were losing their gloss in the dark days of the 1970s. Another model of inspiring curiosity would step into the breach.

OPPENHEIMER AND HIS MUSEUM

Frank Oppenheimer, the charismatic founder of the Exploratorium, wanted to create an institution that would act as a beacon of enjoyment, a celebration of human creativity in an age disillusioned with the fruits of the twentieth-century scientific project. Oppenheimer was a manic trickster figure, whose public image as a playful, uninhibited, childlike genius became integral to his project. As with a few other (always male) figures in kids' science culture in the twentieth century—Don Herbert's Mr. Wizard or A. C. Gilbert of the A. C. Gilbert Company—Oppenheimer's avuncular persona was a promotional tool for his venture. "If the Exploratorium . . . is Wonderland," journalist Ruth Reichl wrote about her visit to the museum in 1978, "then Oppenheimer himself is all the characters rolled into one. It is easy to see him as the Mad Hatter, inviting us all to join in the tea party."[28]

Oppenheimer's famous name and history was catnip to the press, because journalists could set up a resonant opposition between the serious wartime work of the Manhattan Project and the rainbow seventies wonderland of the Exploratorium. While Frank Oppenheimer would never achieve the widespread symbolic fame of his older brother—there are no biopics about Frank and only one published biography—his story had its own potent appeal.[29] Oppenheimer's life story was an irresistible narrative, as his résumé traced a perfect three-act trajectory: success in the world of big science; a period of professional rejection and a pastoral decade of retreat; a resilient reentry with a spectacular and innovative end-of-life project.

Oppenheimer, who was born in 1912, was the younger brother of the more famous J. Robert. The two were raised in a wealthy household in New York City, and their father was an art collector; they each dabbled seriously in the arts (poetry for J. Robert, the flute for Frank) before getting Ph.D.'s in physics and becoming scientists. As scientists, the two operated quite differently, with Robert pursuing a theoretical path and Frank a more physical, hands-on, experimental one—a distinction that would later shape Frank's approach to teaching and museum work.[30] Before the war, while working toward his doctorate at Cal Tech, Oppenheimer researched cosmic rays, constructing balloons that would carry detectors into the sky and traveling internationally to carry out tests. He spent the war years at the Radiation Lab in Berkeley, working for the Manhattan Project on the problem of uranium isotope separation and acting as an all-purpose troubleshooter, tackling problems as they arose.[31] Frank spent time at both Oak Ridge National Laboratory and Los Alamos, where Robert brought him near the end of the war to witness the July 1945 Trinity test. (He often referred to himself, wryly, as "the uncle of the atomic bomb."[32] Along with others in the immediate postwar era who formed the Association of Los Alamos Scientists, Oppenheimer advocated nuclear energy be controlled by national and international regulatory bodies.)

After the war ended, Oppenheimer saw a promising research career cut short when he was investigated by the House Un-American Activities Committee, admitted to having belonged to the Communist Party from 1937 to

1940 (after initially denying his association), and refused to "name names."[33] After being forced to give up his academic position at the University of Minnesota and having his passport revoked, he spent a decade living with his family on a cattle ranch in Colorado, which he bought using money from the sale of an inherited Van Gogh. In Colorado, Oppenheimer taught science at a local high school, taking a hands-on approach and using some of the new National Science Foundation–funded curricula created by physicists and biologists after Sputnik.[34] Upset by teenagers' lack of understanding of the workings of everyday things, he took his students to the junkyard—figuratively, in his inclusion of odd objects in the classroom, and literally, on field trips to the local dump, where they messed around with old auto parts to understand how they worked. The eccentric—and beloved—teacher created a "library" of experimental apparatus for the high school's laboratory.[35] After returning to college teaching in 1959, Oppenheimer found himself dissatisfied with the state of physics as an academic field, calling it careerist, "an intellectual desert."[36] He spent a year in London on a Guggenheim Fellowship, studying three European science museums.[37] Drawing from this research and with a small amount of seed funding, he launched the Exploratorium in 1969.[38]

In the post-Sputnik era, legacy museums struggled to retain funding and prove their relevancy while public rhetoric turned to the insufficiency of science education. Critiques of museum practice centered on efficacy and relevance. As public consensus regarding the need to increase the nation's scientific "manpower" solidified, all older models of science education, including the "exposure" model of science museums (which assumed that visitors would "get" something from displays just by proximity), came under scrutiny. Another class of critiques, which held that science museums were insufficiently political in their approach to discussing science, surfaced in the late sixties and early seventies, as the Exploratorium project gained steam. As the environmental and antiwar movements critiqued scientific alliances with industry and government, science museums came in for their share of scrutiny, in particular the ones who accepted exhibit sponsorship from oil and chemical companies.[39] "I remember one exhibit," Oppenheimer told Ruth Reichl in 1979 in discussing his motivations in starting the Exploratorium, "that had a sign, a big one, that said, 'Steel is for Fun.' Well of course . . . steel is for a lot more than fun."[40] Such exhibits tended to be expensive and hands-off, positive and cheerful about industry at a time when such positivity was a warning sign to a public worried about pollution and overdevelopment.

The Exploratorium's exhibits—which were hands-on, cheap, engaging, and devoid of obvious ideology—quickly became shorthand for the "innovation" that the museum sector seemed to need. Much media coverage of the Exploratorium casts it as a "different kind of museum," contrasting its colorful and anarchic style with the supposed mustiness of the old guard. Directors of other museums soon turned to Oppenheimer for advice and referred to the Exploratorium in their funding applications; the concept of a "hands-on" museum permanently changed the field.[41] The Exploratorium also shunned explicit critiques of the structure and function of science in society in favor of a type of innocent, personal science that was divorced from the outside world. In its exhibit design, it encouraged an elemental, experiential approach to science, inviting the exercise of individual agency and perception.

Although some of the press reporting on the opening of the Exploratorium treated the project as a simple attempt to "make more scientists"— another project of postwar recruitment—Oppenheimer's motivations were much more romantic and idealistic. If the Science Talent Search sought to support a new cohort of young scientists by creating a youth culture that accepted and rewarded their "talent," Oppenheimer wanted to heal a larger culture separated from what he saw as the basic human instinct toward curiosity. He was a prolific author of musings on science, social justice, and education and outlined his philosophy in statements of purpose, speeches, and essays, as well as in his comments to the many reporters who visited the museum or called him while writing stories about science education.

Oppenheimer described his museum's mission as "taking away the harsh view of science."[42] He was convinced that big changes, like the invention of the bomb, were upsetting but that smaller frustrations in the material landscape of everyday life were also currying public negativity. "When you say that there's an anger at science, it's more an anger at the use of technology," he argued. "When you buy a camera, there is never really any good explanation of optics, f-stops or how lenses work—and the same is true for automobiles. The explanations are all junk."[43] The prevailing expectations of consumer and mass culture, he thought, were that people would passively accept new technologies and act as spectators, with little understanding of the gadgets and medicines they used. He believed this incuriosity was fundamentally anti-human, antigenerative; elsewhere, he compared the lack of a drive to understand the world with "celibacy," invoking an image of severe mental sterility.[44]

This, Oppenheimer thought, was a problem not just for individuals but also for the culture at large. Like colleagues in science promotion from earlier in the century, he subscribed to the belief that curious people with a grasp of the scientific method would make better citizens.[45] Writing in 1983, Oppenheimer said that he had the idea for the Exploratorium after feeling that, like his students in Colorado, "too many people . . . had given up the hope of comprehending anything about nature, or even about the everyday gadgets they used or about the history and the workings of the society in which they lived. They had lost the conviction that the world about them is understandable."[46] Oppenheimer recalled the pro-science, pro-technology attitude of the 1930s — the decade of his own young adulthood — with nostalgia. He pointed to the "industrial democracy" of the Soviet Union as an experiment that had made him and his cohort feel like they had the power to change the human world for the better. "There was the sense," he said, "that I could make a difference and my friends could make a difference."[47] The times, he thought, had encouraged not only excitement about advancements in science and technology but also big thinking about the way society could work.

For Oppenheimer, there was a progressive relationship between understanding the way that perception, light, and sound worked and grasping larger concepts in the social world. The Exploratorium's hermetic, safe utopia would generate curious people, who in turn would come to apply their newfound love of thinking to their social worlds. For Oppenheimer and the others who founded the Exploratorium, science's effects on the human psyche tended to create citizens who were more open to connection with each other. In a 1984 speech upon accepting a medal from the American Association of Physics Teachers, Oppenheimer pointed to what he called "the sentimental fruits of science": "What we discover about nature plays a deep and basic role in the way we think about ourselves, about other individuals, and about society as a whole."[48] Oppenheimer believed that explorations in science were particularly important in the fractious times of the Cold War because they could help people become more open to recognizing the symmetries of human experience. In fact, such broad-minded people might even be better for the world than scientists would be. When asked whether he thought there should be more scientists in the United States, Oppenheimer replied: "It's a very enjoyable life. So it'd be a good thing. I don't know if there'd be a need for them. I think scientists that are rounded out a little bit are not such a bad lot, if they're just too narrow they're pretty awful."[49] For Oppenheimer, the *mindset* that science play endowed was more important than the increase of technical acumen.

Exploring the factors that could contribute to the loss of curiosity he felt afflicted the young people of the 1960s and 1970s, Oppenheimer thought that fear—of physical dangers or of appearing or feeling "stupid"—was a major culprit: "As a child you are taught many fears: not to play with wall plugs, not to talk to strangers, not to stick your head out of a car and feel the wind. You are told, 'curiosity killed the cat.'"[50] Fear of academic failure was one of the demons that Oppenheimer hoped to banish. In 1970, Oppenheimer wrote (in a phrase he often repeated): "No one ever flunks a museum, one museum is not a pre-requisite for the next. People do not list the museums they have attended on a job application form. Museums are thus free of many of the tensions which can make education unbearable and ineffective in the schools."[51] Older people, too, still felt the aftereffects of that fear; Oppenheimer wanted the Exploratorium to act as a big reset button that would wash anxieties surrounding science from the minds of its visitors.

The elimination of these "tensions" was foremost in his mind when he planned the exhibits and physical plant for the Exploratorium. In writing a document on exhibit conception and design for fellow museum professionals, Oppenheimer reiterated that he wanted visitors not to fret about what they might have been "supposed" to get out of their time on the floor: "The atmosphere of the museum must be adjusted in such a way that people are not afraid about missing or not understanding something."[52] Exhibits, he wrote, should not be frustrating, and museum planners should not be overly concerned about the lessons they should impart. "Some of the things that we put out there are there just because they feel nice," he said. "For example, we have one exhibit that is simply a big ball-bearing. Next to it is a sign that says, 'Some machinery feels nice.'"[53] To reporters who might perceive the museum's purposefully noncoercive layout as random, Oppenheimer used his metaphor of a forest, "a woods of natural phenomena that were organized and selected in some way so that people could take many constructive paths."[54]

Some educators observed the interaction between children and this "forest" of a museum with a critical eye. "The [explanatory] signs often are ignored," a writer for *American Education* observed in 1978. "A case in point: A ball is bouncing crazily over a stream of air. Alan and John [two young visitors] can't resist the temptation to seize it, pull it out of the air stream, and create a new game. 'Hey, Alan, let's see if we can throw it through the air stream.' The children may not discover for a long time that the ball is demonstrating Bernoulli's Principle and that they could learn about the lift on an airplane's wing by observing the pull on the ball."[55] Oppenheimer refused to be concerned about stories like this one, choosing to celebrate children's good feelings about exploration rather than mourn missed opportunities. After decades of instrumentalist science promotion, the stance was a radical one.

A CHILDISH SPACE

Oppenheimer's persona and philosophy, as reflected in his museum, defined a politics of individual scientific liberation using the visual language of childhood play. This move drew upon new understandings of the nature of childhood in the 1970s. The vogue for creativity research in the postwar period had found particular traction in midcentury children's culture, as architects, toymakers, and interior designers incorporated its tenets into playscapes and educational facilities.[56] On a much more radical level, the liberation movements of the 1960s inspired early-1970s writings by children's liberationists, who called for the loosening of the grip of the nuclear family and the school on children's lives. Children's liberationists wanted child-centered education—the idealistic dream of the Progressive Era—to be a real and available option for young students, even those without the resources to go to private schools. Such activists called for the "island" of twentieth-century childhood, a segregated space maintained for the supposed protection of children, to be dismantled and for children's and adults' worlds to be commingled.[57] They wanted children to have rights: passive rights like health care and education, but also active rights to choose the circumstances of their lives, by voting, working, or leaving their parents' homes to live elsewhere.[58] While Oppenheimer was not a children's liberationist, in any explicit sense, the Exploratorium's policies and physical plant answered many of these concerns. (He celebrated the news when he found that kids were "playing hooky" from school to come to the Exploratorium, aligning himself emphatically with freedom over law and the expectations of authority.)[59] Oppenheimer certainly aligned himself with a romantic valuation of childhood creativity, which celebrated permission and expressiveness, through his institution's attempt to redefine what it meant to "do science."

A newspaper reporter once described the Exploratorium director as possessing "a boy's zest and a scientist's dedication," and Oppenheimer often represented himself as embodying the ideal balance between these qualities.[60] In his writings, he remembered childhood episodes of exploration, including a story in which the young Frank traversed the house with a bottle, inserting a tiny bit of every spice, chemical, and drug he could find, and another in

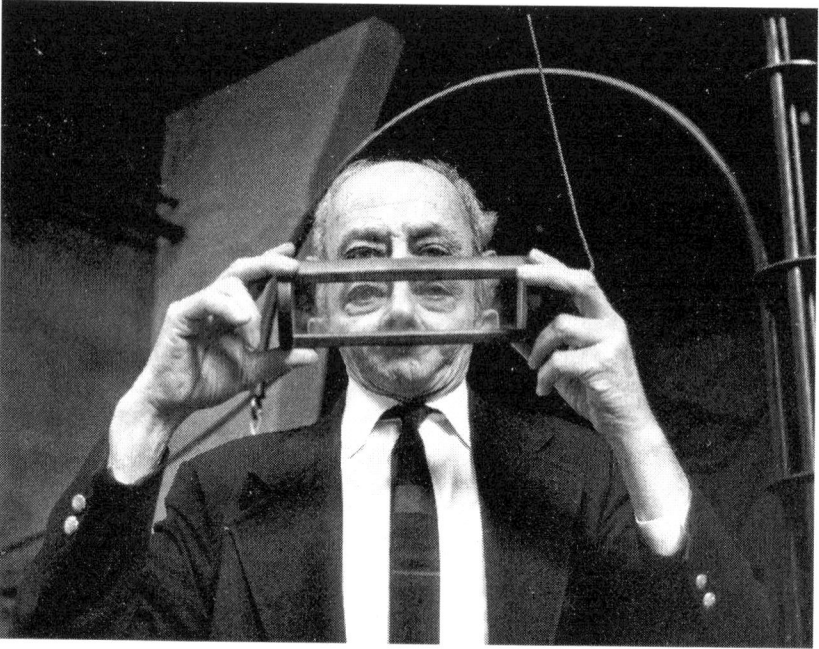

"Frank's Unique View of the World." KC Cole, © Exploratorium, *www.exploratorium.edu.*

which he ran his finger along the edge of a circular knife in a butcher shop ("I bled profusely").[61] Oppenheimer described himself as a large child, still playing whenever he could; in "Adult Play," a 1980 article for the Exploratorium's magazine, Oppenheimer admitted, "There are times when driving that I keep time to radio music with the accelerator and the brake to produce a quite remarkable motion of the car."[62] K. C. Cole, a journalist and friend who wrote a biography of Oppenheimer and the Exploratorium, described him as "Tom Sawyer in a business suit": "He fidgeted endlessly, fiddling with small objects he kept in his desk or his pockets: a slide rule, a top, a magnifying glass, a pocket spectroscope. He smoked nonstop, and on more than one occasion set himself on fire by putting out butts in his pockets."[63] Oppenheimer's eccentricity and subversion of normative adult behavior made him an icon of the kind of curiosity and creativity that he was trying to cultivate in the museum's visitors.

The museum characterized its target audience as age neutral; all who entered, its publicity argued, would become as children, transformed by the space. In a reversal of the concept behind the early twentieth-century Brooklyn Children's Museum—a place that was the "children's own"—the Exploratorium was a space in which age was equalized, rather than celebrated. The Exploratorium, which charged no admission until 1981,[64] understood itself as democratic, "a favorite playground of both schoolchildren and nuclear physicists, of artists and little old ladies."[65] (Oppenheimer once compared the makeup of the museum's visitors to "the crowd at a public beach.")[66] Oppenheimer believed that the pursuit of understanding had the power to render all young again: "Our exhibits encourage people to ask, and then answer for themselves, the question: 'I wonder what would happen if I did this or that?' Some people say that the asking of this question makes children out of adults."[67] He told a reporter a story about a group of "elderly folk from a rest home" who had visited; "one woman had just used a giant balloon which allows people to communicate with others across the room. Her face lit up. 'This is the most wonderful place I've ever been!' she exclaimed. 'I had to wait 70 years to find a place like this!'"[68] The magical equalizing properties of the museum often found visual representation in its publicity. A popular exhibit for years was the "Distorted Room," which altered the proportions of people inside, leaving children looking bigger than parents and teachers. The exhibit, which visually referenced "Alice in Wonderland," made for a great photo opportunity.[69]

Perception was a central theme of the museum, and one that enforced the ethos of democratic inquiry. Exploratorium exhibits, reminiscent of nineteenth-century "philosophical toys," played with visual illusions, perspective, optics, physics of sound, music, relative motion, and tactile perception.[70] Artists were often invited to contribute exhibits. Photographs of child visitors during the 1970s highlighted the colorful nature of the Exploratorium's interior spaces, often featuring a silhouetted figure of a child against a rainbow backdrop. *The Sun Painting*, a centerpiece exhibit designed by local artist Bob Miller, was prominently displayed at the museum's entrance, showcasing a changing mix of colors on a screen, which came in and out of focus depending on the sunniness of the day.[71] Reflection was another prominent theme. As a press release described the space: "[The Exploratorium] is a land of a thousand mirrors: curved mirrors, inside-out mirrors, kaleidoscopes by the score."[72] The museum had a visual and conceptual connection to the psychedelic movement, which believed in the liberatory potential of changed perceptions; across the Bay, physicists at Berkeley experimented with mysticism and Eastern spirituality while exploring the roots of quantum theory.[73]

From an article in *Smithsonian* magazine, September 1978: "It's OK to Touch at the New Hands-On Exhibits," by Sherwood Davidson Kohn, photography by Christopher Springmann and Jonathan Atkin. *Smithsonian* magazine.

The choice of the senses and perception was meant to encourage the growth of a sense of agency in the museum's visitors. Oppenheimer wrote that the museum focused on perception because "it has the virtue of encouraging even lay people to argue and ask meaningful questions about a subject." This was particularly important for young visitors, he wrote: "Elementary school students never feel free to argue about physics with an instructor or among themselves. They might ask questions, but they don't say 'You're wrong.' But with perception, this sense of back and forth argument can happen because there are so many variations in the way people can see and hear."[74] Rather than impart a given body of knowledge, the Exploratorium wanted to teach that knowledge could sometimes be contingent. "Reality" wasn't the same for every person—a body of facts ready for access—but its contingency did not negate the importance of trying to understand. Moreover, the very fact of that contingency leveled the playing field between children and adults, making intergenerational discussions possible.

Rather than emphasize the wondrousness of past scientific achievements or tie these achievements to the life stories of successful individuals, the Exploratorium wanted to present science as a wide-open field, equally available to all comers. Oppenheimer wrote, "The Exploratorium is not designed to glorify anything. We have not built exhibits whose primary message is, 'Wasn't somebody else clever,' or 'Hasn't someone done a great service to mankind and the American way of life.'"[75] By diminishing these past accomplishments, the museum would put the visitor in primary relationship with the natural phenomena being explored. "We must not glorify the achievements of scientists, artists, engineers, or businessmen," Oppenheimer wrote. "We must make it possible for visitors to feel that *they* are the clever and perceptive ones, not the scientists and engineers."[76] The Exploratorium equalized levels of expertise, as well as age; another reporter observed a group of scientists working on a steam generator and wrote, "They have all the enthusiasm of youngsters trying to get their coaster to roll down a hill. 'This place brings out the kid in all of us,' says one of the scientists, standing atop the 43,000-pound generator. '[Scientists] go years and years to college and then stand around and shout: 'Make it work, spin and light up.'"[77]

In order to decentralize authority still further, Oppenheimer hired high school students to act as "Explainers," mingling with the visitors and talking about the exhibits and their effects. Oppenheimer got the idea from his visits to the Palais de la Découverte, a science museum in Paris, where young people demonstrated apparatus and answered questions.[78] The youth of the Explainers encouraged a flat playing field between docent and visitor. "One of the troubles with teaching science in general is that this collective right answer of the scientist dominates the field so that pupils never argue with instructors, or even with each other about the validity of what's being said," Oppenheimer said in a 1981 interview for a *NOVA* documentary about the Exploratorium.[79] Explainers, who were not science prodigies and were often learning about the exhibits along with the visitors, were open to discussion. The Explainer program resulted in a learning experience for the teenagers involved, as well as for the museum guests. Oppenheimer, who thought of teaching as one of the highest human pleasures, repeatedly argued that instruction was the best way to learn any subject. While adult teachers might try to enlist students in peer learning in classrooms, he wrote, this was not the right setting for the endeavor. The Exploratorium, on the other hand, was forgiving enough to allow for the practice: "Young students can teach in a museum by demonstrating or fabricating particular exhibits, although they might feel incompetent or embarrassed to do so in a classroom."[80]

The Exploratorium made a special effort to hire Explainers, who were generally juniors and seniors in high school, from diverse backgrounds, both to provide opportunity and to bind the museum closer to the community. One newspaper article reported that "the Explainers form an important bond between the institution and the community . . . their presence has attracted neighbors and friends who might not have visited the museum."[81] The job required no particular experience with science, technology, engineering, and mathematics (STEM). Students were interviewed and assessed on the way they interacted with the exhibits, and their interpersonal abilities and public-speaking potential. Importantly for the museum's efforts at inclusiveness, Explainers were paid.[82]

If adults became children in the Exploratorium and teenagers became teachers, children's childishness was emphatically allowed. Oppenheimer called the museum "manifestly noncoercive."[83] Students were not required to line up and experience the museum's exhibits in order; the Exploratorium self-consciously rejected regimentation. "It is impossible to lead a group through [the Exploratorium] on a guided tour," Oppenheimer wrote in 1972. "If one starts off with a group, one soon finds oneself alone, other people having stayed behind to play with or investigate one or another of the displays of the intended tour."[84] A story Oppenheimer told interviewers for the *NOVA* documentary shows how much the Exploratorium's ethos of distributed exploration inverted expected pedagogical relationships. "Long ago," he said, "I met a parent who brought his young boy [to the museum], and the boy kept running off, and the parents said 'Come, I want to show you this,' and the boy disappeared, and then a little while later the boy came running up and said 'Hey Dad, come look at this.' And the father actually said 'You're disgusting, I brought you here because I know this stuff, I want to show it to you. You're not supposed to be showing stuff to me!'"[85] Oppenheimer's intention in telling this story may have been to point out how restrictive some children's relationships with their parents were and what a negative effect this could have on a child's curiosity. But the anecdote makes another argument as well, showing a case in which the Exploratorium had the power to reverse this effect — if only for an afternoon.

Oppenheimer and others at the Exploratorium recognized that its stance on the freedom of physical movement was one that was strange to other adults engaged in planning similar public spaces, and they prided themselves on their extreme tolerance of motion and noise. In a 1983 interview Oppenheimer described a conversation with architects who tried to figure out ways to keep children from running in the museum. "I had a difficult time convincing the architects that they weren't really doing any harm," he said. "They hardly ever ran into anybody, but they appeared to be a little bit out of control and it worried the architects to see children behaving naturally. I think it's quite wonderful that we don't mind losing some control."[86] Writers often noted the physicality of young people's interactions with the museum space; a reporter for *Mosaic* magazine noted that Oppenheimer's one rule was the prohibition on riding a bicycle between exhibits and that this was "less to protect the exhibits than to avoid collisions between the kids darting from one to another."[87] ("The only rule is, no bicycles," was a charming trope repeated across many articles and interviews about the museum.) Writing in a magazine for museum professionals in 1972, Evelyn Shaw, the curator of life sciences at the Exploratorium, acknowledged that participation could create problems for people who worked at museums: "Turn the dial, manipulate, change the controls, do it yourself. These phrases, associated with participatory exhibits are anathema to many science museums. They spell trouble for the exhibition departments, for the maintenance staff. They imply problems, seemingly endless adjustments, and exhibits frequently shrouded in 'out-of-order' signs. Participatory exhibits often create severe headaches. Are they worth the migraines?"[88] The answer, Shaw went on to say, was definitely yes. The proof, for the Exploratorium, was in the attachment children felt to the place.

Images used in magazine and newspaper publicity often included pictures of children — like Amy Carter — spinning, running, or playing. Many visitors noticed the noise level. "There are . . . as far as the eye can see, exhibits which twinkle, beep, wail, pound, flash, receive and bounce back light, and cause shrieks, laughter, and pure enjoyment. The din of the crowd drowns out the fog horns signaling the movement of ships in nearby San Francisco Bay."[89] Unconsciously echoing language used by the Brooklyn Children's Museum staff fifty years before, K. C. Cole wrote that the Exploratorium's freedom even had the power to change children's behavior. "The Exploratorium belies the myth that small children have short attention spans," Cole wrote. "They linger

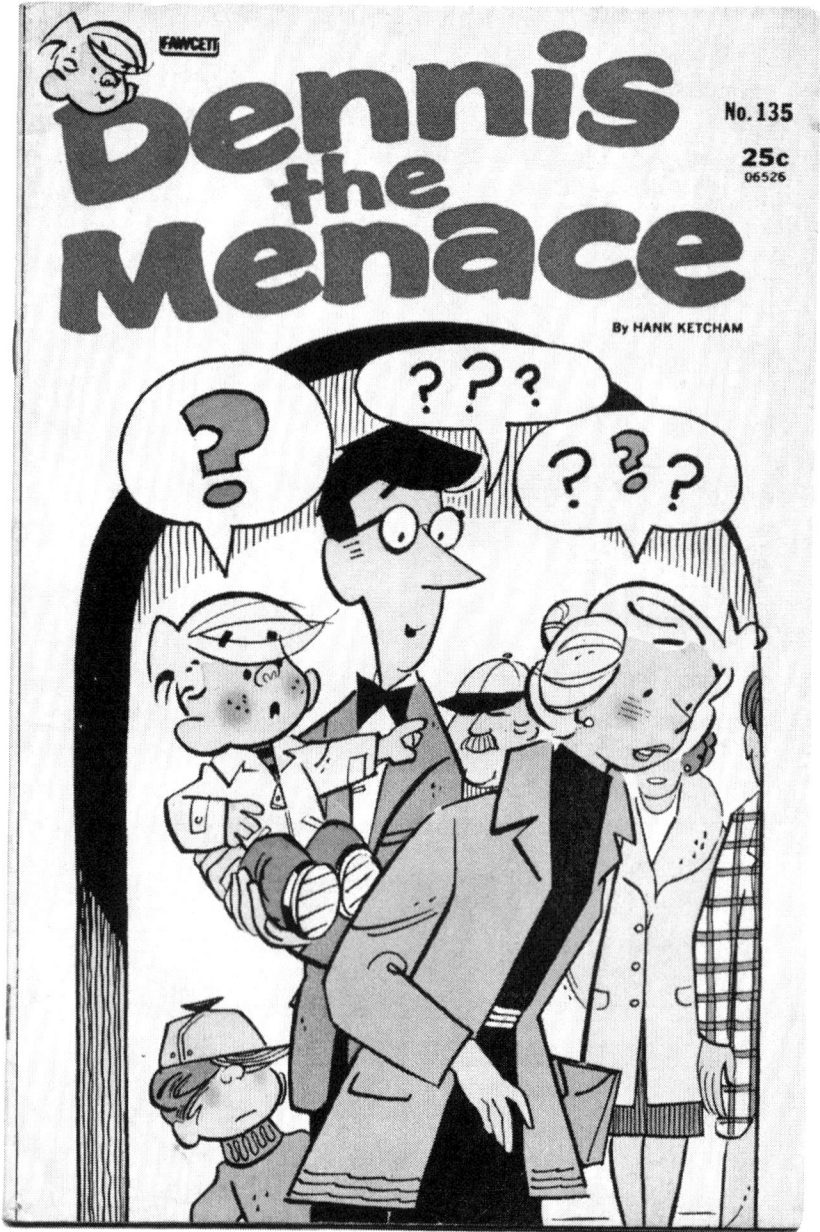

Cover of the *Dennis the Menace* comic book, no. 135. Fawcett Publications.

for long periods, standing on tiptoe, eyes wide, noses pressed to glass, oohing and aahing as if it were Christmas morning. Stools and steps and child-sized signs are everywhere. 'You never hear a baby cry,' says a staff member. It's the ultimate distraction."[90]

One particular pop-cultural appearance of the Exploratorium shows how its ethos of absolute freedom and positivity aligned with other ideas about creative childhood in the postwar period. In a 1974 issue of the *Dennis the Menace* comic book, Dennis, the ultimate icon of postwar permissive parenting, and his long-suffering parents, Hank and Alice, go on a road trip and visit the Exploratorium.[91] Hank, Alice, and Dennis join together in being mystified and intrigued by the museum exhibits; their faces, on the cover of the comic book, are surrounded by swarms of question marks. Dennis, living up to his nickname, breaks rules and jumps on a piece of museum equipment before he is supposed to; his father chastises him, but the docent is gently accepting, musing, "This isn't the way it's supposed to work . . . but you *have* made a different design."[92] In another sequence, Dennis watches as his parents enter the "Distorted Room," which makes his mother look larger than his father—another Wonderland-style inversion of power. Dennis's amazement carries him from room to room, his legs shaking with excitement. Dennis the "Menace" has found a science museum that matches his inquisitiveness and tendency toward chaos.

CONCLUSION

Frank Oppenheimer died in 1985. In *Leonardo*, a journal for artists with STEM interests that was founded in 1968 (a year before its fellow traveler, the Exploratorium), San Francisco science writer David Perlman eulogized the educator as "a gentle, courtly, humorous and profoundly humanistic scientist."[93] The museum is still open today, in a different location, and with a somewhat more professional feel; admission will now cost you quite a bit of money.[94]

The Exploratorium in the 1970s and 1980s was a charming, charismatic place, run by a charming, charismatic man. It exerted a special pull on the imaginations of journalists looking for a human interest story, artists with broad-minded approaches to collaboration, and museum directors looking for a way to differentiate themselves and attract funding. The idea of the museum was singular—nothing like it had existed before. While its ideology was deeply romantic, almost radical, the story of its shoestring origin had a rags-to-riches, upstart appeal that was classically American. Oppenheimer's deep credibility as a scientist with one of the most famous scientific names in the United States grounded the place, while his rebirth as a whimsical educator made for great copy.

Yet the fundamental reason why the Exploratorium story resonated so much in the 1970s had to do with the recouping of a utopian world of science play. Whether the visitors were adults or children, their unleashing of a proper sense of wonder, the Exploratorium promised, would bring them back to the world of childhood. To accomplish this trick, the world of the Exploratorium had to remain apart from the big questions about science and technology that troubled the 1970s and 1980s. Childhood "curiosity" was a perfect balm for those ills.

Conclusion

Looking Closer at "Kids Are Little Scientists"

Scholars working in childhood studies are quite often confronted by what I think of as the "cuteness problem." American beliefs about childhood—"boys like trucks"; "kids are so innocent"; "children love candy"—seem ingrained enough to feel biological, exempt from cultural analysis; they are deeply appealing. The idea that "kids are little scientists" is one of these indelible tropes, a bit of twenty-first-century folk knowledge that pleases many people. Much of my own initial interest in this topic came from the inherent cuteness of the historical archive—the covers of chemistry sets, the space cadets in their helmets, the earnest "research reports" from science clubs. So I cannot much blame people who come up to me after I give a talk about chemistry sets and want to tell me anecdotes about their uncles blowing up their basements; it can be hard, indeed, to regard this history with an analytic eye.

Yet, the shapes that this particular trope has taken over the years have been culturally derived and have had real influence. The repeated alarms raised by politicians over the sufficiency of the science, technology, engineering, and mathematics (STEM) workforce since World War II are, as researcher Michael S. Teitelbaum has shown, not only overblown but also counterproductive, leading to a cycle of booms and busts that have made careers in science and engineering in the United States an unstable proposition.[1] Writing about Teitelbaum's work for an industry website aimed at engineers and technologists, science journalist Robert Charette points to the fact that many STEM graduates do not have STEM-related jobs, and vice versa. Charette asks: "If many STEM jobs can be filled by people who don't have STEM degrees, then why the big push to get more students to pursue STEM?"[2]

Why, indeed? Charette argues that for companies employing STEM workers and for universities looking for funding for STEM programs, the half-century-old belief in a STEM shortage is economically advantageous. But I think that the belief is also easy to sell, culturally speaking, because it is a critique of young people's choices — and, on a more deeply anxious level, a critique of the culture that created those young people. If "kids are little scientists" but never end up becoming big scientists, why is that? The answer, since the postwar period, has often taken the form of a jeremiad. Whether it is *Life* blaming a fun-loving teenage culture or Robert Heinlein blaming (female) children's book editors and teachers, or Frank Oppenheimer blaming boring schools that induce performance anxiety, theorizing about kids' disinterest in science is also theorizing about the conditions that influence children's worlds.

Quite often, the assignment of blame carries within it a significant dose of nostalgia. Lately, laments over a lack of scientific interest come bundled with the widespread critique of the "overprotection" of children.[3] Perhaps the most nostalgic reminiscences about an unbounded material culture that was once available to children come from scientists and science promoters mourning the apparent loss to scientific experimentation through a more regulated home environment. These arguments are historical, pressing a common understanding of "childhoods past" into service to make an argument about the trajectory of society as a whole. As early as 1980, *New York Times* columnist Malcolm W. Browne was eulogizing the "magic," "malodorous and occasionally alarming" home chemistry lab. "The careers of many . . . leading scientists were shaped by home experimentation, in which there was always time to grope for understanding at one's own pace and in one's own way," Browne wrote. "Mounting social and legal pressures to make life safe and supervised have incidentally robbed science of some of the magic with which it once lured young people."[4] More recently, *Wired* science writer Steve Silberman wrote a similar piece, echoing Browne's image of science's lost chance to work its seduction: "The lure of do-it-yourself chemistry has always been the most potent recruiting tool science has to offer."[5]

Many of these nostalgic diagnoses implicitly tie a lost "love of science" with cultural keywords associated with boyhood: "messiness," "mischief," "danger." In 2013, following the publication of a study showing scientists' hiring bias against young female job applicants, journalist Eileen Pollack (who had, herself, received an undergraduate physics degree and then left STEM behind) asked, "What could still be keeping women out of the STEM fields?"[6] Pollack's investigation found that the culture of high school and college classrooms contributed to this problem; she pointed to a lack of encouragement from teachers, a sense that STEM knowhow made girls unattractive, and an expectation of competition and overwork.

The history of the "kids are little scientists" trope shows that there are also answers to this question to be found in cultural expectations about younger children. If the wireless boy, the basement experimenter, and the space cadet are still the symbolic figures that convey "science kid," then "science kid" is not gender neutral. The "boy genius" is a trickster, a mess, a force of nature; he embodies transgression against adult desires. Today, many popular science programs and materials for kids are marketed as "horrible" or "disgusting." The United Kingdom–based book series *Horrible Science*, by Nick Arnold, published by Scholastic, is a prominent example, with such titles as *Deadly Diseases, Blood, Bones, and Body Bits, Chemical Chaos, Explosive Experiments,* and *Suffering Scientists.*[7] As children's literature scholar Annette Wannamaker points out, contemporary popular culture aimed at boys uses the repulsive as a way to mark itself as "boyish"; enjoyment of the abject creates a boundary between kid and adult and between boy and girl.[8] "Disgusting" science, nominally aimed at all children, is (let's face it) for boys.

The *boyishness* of science is also deeply ingrained in the words people use to discuss the feeling of scientific discovery. After Caltech astrophysicist Shrinivas Kulkarni made a reference to scientists as "boys with toys" on a National Public Radio talk show in May 2015, female scientists took to Twitter to share images of themselves with their own "toys": Mars rovers, computers, lab equipment.[9] The women's reactions challenged a powerful icon — the male scientist regressing to boyhood through the force of interest — and made it clear that female scientists, too, experienced that joy of play. On a subtler level, this challenge to a figure of speech showed just how ingrained our idea of science's boyishness can be. "Girls with toys" neither rhymes nor evokes the same vision of free, absorptive childhood play. Perhaps this is because, historically, girls' play has not been accorded the same time and space as that of boys and certainly has not been regarded so adoringly.

Gender implications aside, "kids are little scientists" can misrepresent the nature of scientific practice. In a 2012 blog post, science communicator and educator Marie-Claire Shanahan tried to explain why that trope bothered her. "Describing [children] as scientists is a wonderful and rich analogy," she wrote. "The problem is that like all analogies, it breaks down, and this one breaks down especially clearly when it's applied specifically to science learning. Child-as-natural-scientist arguments tend to equate curiosity and exploration with the expert practice of science, something that under-plays and devalues the difficult and complex world of scientists. . . . Science isn't just a grown up version of a child's curiosity. While kids have the fertile beginnings, becoming a scientist requires that they learn and skillfully practice many abstract skills that are far from intuitive."[10] Science, Shanahan and others point out, can be complex, boring, bureaucratic; experiments you work on for years can fail without warning, and you may rarely get that feeling of excitement and "thrill" that chemistry sets advertise. I once saw an advertisement for a summer camp for elementary school children, called Science Explorers, that led with the tag line, "If it's not fun, we're not doing it!" The expectation that science should be pure joy, which is often proffered to counteract what promoters perceive as negative cultural baggage, offers a confusing image of what it means to do real scientific work.

"Kids are natural scientists" also excludes a significant population from the love of science: adults. Ariel Waldman, a science promoter who works with NASA and invented an all-ages, worldwide program called Science Hack Day, wrote in 2014 that she is often asked about her interest in space as a young person. "The interviewer is throwing me what they believe to be a softball question," she wrote, "assuming with almost certainty that I'll have a quaint heart-warming story about how I now have achieved what I always dreamed about as a kid." But Waldman never loved space or science as a kid; she loved art and design. She got a job at NASA as an adult and, as she put it, "awoke to my obsession with space exploration." Waldman uses her story to plead for the adults who have not yet been "converted" to loving science, pointing out that science promotion for adults who last studied a STEM subject in high school or college is "a complete void." Echoing Frank Oppenheimer, Waldman argues that "having adults play with science, despite lack of understanding or knowledge, is the most empowering form of science engagement. . . . Science shifts from being Fort Knox to being just another available material you can manipulate."[11]

In the late 1980s, I was part of a middle-school enrichment program that asked us to build a mousetrap. The project was presented to us as the ultimate in fun: Take a paper bag of humble objects home with you, pour them out on a table, and mess around until you figure out how they should fit together in mousetrap shape. I hated it. I was a reader and a researcher, in love with narrative; I liked writing things and putting on plays with my friends. I couldn't imagine anything that worked would ever come out of the motley pile of paper clips and corks, and I whined at my parents, making their lives insufferable. I dropped out of the program that week, ashamed of myself and annoyed at the world.

My anecdote of failure, on its own, proves little about my aptitude for STEM. Perhaps I would have been better off in a gardening club, spotting birds, or playing with a telescope; perhaps this utter frustration of the mousetrap didn't mean that science was not for me. But what I remember most from the experience — and what I think is significant to include in a study about the way a particular cultural ideology moves through generations — was my frustration at myself. "I *should* like this," I thought. "What's wrong with me, that I don't like this?"

Maybe now I should try again?

Notes

Abbreviations

BCM Brooklyn Children's Museum Archives, Brooklyn Children's Museum, Brooklyn, N.Y.

CHF Chemical Heritage Foundation Archives, Philadelphia, Pa.

ER Exploratorium Records, BANC MSS 87/148 c, Bancroft Library, University of California, Berkeley

GN Gaylord Nelson Archives, Wisconsin Historical Society, http://nelsonearthday.net/index.htm.

HA Robert A. Heinlein Archives, Heinlein Prize Trust, University of California, Santa Cruz

PP G. Edward Pendray Papers, Princeton University Archives, Princeton, N.J.

SA Smithsonian Institution Archives, Record Unit 7091, Washington, D.C.

SM Archives, Strong National Museum of Play, Rochester, N.Y.

Introduction

1. Barack Obama, "Remarks by the President at the White House Science Fair," February 7, 2012, http://www.whitehouse.gov/photos-and-video/video/2012/02/07/president-obama-speaks-white-house-science-fair#transcript (accessed February 19, 2015).

2. Tolley, *Science Education of American Girls*.

3. Sánchez-Eppler, *Dependent States*; Castañeda, *Figurations*; Sammond, *Babes in Tomorrowland*; Stephens, *Children and the Politics of Culture*.

4. See Gopnik, *Philosophical Baby*; Gopnik, *Scientist in the Crib*; Jonah Lehrer, "Every Child Is a Scientist," *Wired Science*, September 28, 2011, http://www.wired.com/wired science/2011/09/little-kids-are-natural-scientists/ (accessed August 23, 2013); William

Phillips, "Why Are Scientists Most Like Children?," *World Science Festival*, June 2010, http://worldsciencefestival.com/videos/why_are_scientists_most_like_children (accessed August 23, 2013); Tim Seldin, "Children Are Little Scientists: Encouraging Discovery Plan," *Child Development Institute*, http://childdevelopmentinfo.com/child-development/play-work-of-children/children-little-scientists/ (accessed August 23, 2013); Sarah D. Sparks, "Study: Young Children Explore as Scientists Do," *Education Week, Inside School Research*, September 27, 2012, http://blogs.edweek.org/edweek/inside-school-research/2012/09/study_young_children_explore_a.html?cmp=SOC-SHR-FB (accessed August 23, 2013); Ashley P. Taylor, "Kids Play the Way Scientists Work," *80beats*, October 2, 2012, http://blogs.discovermagazine.com/80beats/2012/10/02/kids-play-the-way-scientists-work/#.UhdXqRmtJbI (accessed August 23, 2013).

5. Jad Abumrad, "Numbers," *Radiolab* (WNYC, n.d.), http://www.radiolab.org/story/91698-innate-numbers/ (accessed August 23, 2013); Ira Glass, "Kid Logic," *This American Life* (WBEZ, June 22, 2001), http://www.thisamericanlife.org/radio-archives/episode/188/kid-logic (accessed August 23, 2013).

6. James and Prout, "Re-presenting Childhood," 236.

7. See Bederman, "'Teaching Our Sons to Do What We Have Been Teaching the Savages to Avoid': G. Stanley Hall, Racial Recapitulation, and the Neurasthenic Paradox," chap. 3 in *Manliness and Civilization*.

8. Gould, *Ontogeny and Phylogeny*, 119.

9. Secord, "Knowledge in Transit," 662.

10. Fyfe and Lightman, "Science in the Marketplace"; Keene, *Science in Wonderland*; Kohlstedt, "Parlors, Primers, and Public Schooling"; O'Connor, *The Earth on Show*; Secord, *Victorian Sensation*.

11. Angier, *The Canon*, 1–3.

12. Zelizer, *Pricing the Priceless Child*, 11.

13. Cook, *The Commodification of Childhood*; Jacobson, *Raising Consumers*.

14. See, for example, the continuing popularity of James Dobson's books on discipline. Dobson, *Dare to Discipline*, and *The New Dare to Discipline*.

15. Jenks, *Childhood*, 70–80.

16. Walkerdine, *The Mastery of Reason*, 190.

17. Ibid., 5–6.

18. Sharon Stephens, "Introduction: Children and the Politics of Culture in 'Late Capitalism,'" in *Children and the Politics of Culture*, 6.

19. Gutman and Coninck-Smith, "Introduction," 5.

20. Kohlstedt, "Parlors, Primers, and Public Schooling," 439.

21. Rauch, "A World of Faith on a Foundation of Science."

22. Edward L. Thorndike, "Sentimentality in Science Teaching." *Educational Review* 17 (1899): 57–64.

23. By 1899, humane educators and authors had been trying to reach adults and children through fiction and curricula for two or three decades. Unti and DeRosa, "Humane Education Past, Present, and Future"; Mason, *Civilized Creatures*.

24. Kidd, *Making American Boys.*

25. Cross, "The Cute Child and Modern American Parenting," 30.

26. See "I'm Bored!," in Stearns, *Anxious Parents.*

27. Benedict, *Curiosity,* 5.

28. Dewey, *How We Think,* 32.

29. Ibid., 33.

30. See, for example, Sánchez-Eppler, "Castaways"; Hines, "Review."

31. Mickenberg, *Learning from the Left.* See, in particular, chap. 6, "The Tools of Science: Dialectics and Children's Literature," and the epilogue, "Transforming an 'All-White World.'"

32. Kohlstedt, *Teaching Children Science*; Puaca, *Searching for Scientific Womanpower*; Rossiter, *Women Scientists in America: Struggles and Strategies*; Rossiter, *Women Scientists in America: Before Affirmative Action*; Terzian, "Science World"; Tolley, *The Science Education of American Girls.*

33. Kohlstedt, "Parlors, Primers, and Public Schooling," 437.

Chapter 1

1. Hein, "Progressive Education and Museum Education."

2. Anna Billings Gallup, "The Children's Museum as Educator," *Popular Science Monthly,* April 1908.

3. Late nineteenth- and early twentieth-century childhood, Anna Redcay points out in her work on child authors and *St. Nicholas Magazine,* was perceived as paradoxically natural and in need of adult husbandry: "As childhood became a protected, privileged state, those traits generally considered natural to the child were no longer always perceived as being attainable without careful adult guidance." Redcay, "'Live to Learn and Learn to Live,'" 58.

4. "The Children's Museum," *The Year-Book of the Brooklyn Institute of Arts and Sciences* 12 (1899–1900): 12.

5. Barrow, "The Specimen Dealer," 494; Keeney, *The Botanizers*; Kohler, *All Creatures*; Leach, *Butterfly People.*

6. Kohlstedt, *Teaching Children Science,* 13–20; Armitage, "Bird Day for Kids"; Armitage, *The Nature Study Movement.*

7. Theodore Roosevelt, "My Life as a Naturalist," *American Museum Journal* 18, no. 5 (May 1918): 324.

8. Barrow, "The Specimen Dealer," 523; Price, *Flight Maps.*

9. Conn, *Museums and American Intellectual Life*; Kammen, *Mystic Chords of Memory*; Harris, *Cultural Excursions*; Rader and Cain, "From Natural History to Science."

10. Schlereth, *Cultural History and Material Culture.* Schlereth points out that the children's museum was a Progressive Era idea that became truly successful in the postwar era. He writes that in 1941, only eight children's museums in the United States had their own facilities, whereas by 1985, more than fifty institutions could claim that status.

11. Kohlstedt, *Teaching Children Science,* 3.

12. Ibid., 67.

13. Anne Lloyd, "In the Children's Museum," *Children's Museum News*, May 1920.

14. Quoted in Popkewitz, *Inventing the Modern Self and John Dewey*, 32.

15. Glover and Dewey, *Children of the New Day*, 11.

16. Alexander, *The Museum in America*, 53.

17. Gallup, "The Children's Museum as Educator," 371.

18. Caroline M. Worth, "The Children's Museum," *Childhood* 1, no. 4 (June 1922): 11.

19. Sydney Reid, "The Children's Wonder-House," *Independent*, January 4, 1912.

20. See historian John R. Gillis's discussion of the word "islanding," following German sociologists Helmut Zeiher and Helga Zeiher, in Gillis, "The Islanding of Children," 318.

21. Reid, "The Children's Wonder-House," 33.

22. Elizabeth E. Scantlebury, "Spokane Should Have a Museum for Children," n.d., BCM.

23. "From the Children's Point of View," *Children's Museum News*, February 1919.

24. "Library Notes," *Children's Museum News*, December 1926.

25. Pindar, *Nature Treasures of New York City*.

26. "One Boy's Achievement," *Children's Museum News*, October 1919.

27. Chinn, *Inventing Modern Adolescence*; Klapper, *Small Strangers*; Muchnick, "Nature's Republic," http://gradworks.umi.com/34/40/3440577.html (accessed June 26, 2015).

28. Onion, "Picturing Nature and Childhood."

29. Reid, "The Children's Wonder-House."

30. Charles, "Bedford-Stuyvesant."

31. "The Tale of a Microscope," *Children's Museum News*, March 1932. Sydney Reid also mentions children arriving at the museum "dragging unwilling nurses" ("The Children's Wonder-House," 33).

32. "The Children's Museum," *The Year-Book of the Brooklyn Institute of Arts and Sciences* 12 (1899–1900): 249.

33. "The Children's Museum," *Junior Eagle*, February 22, 1914.

34. Klapper, *Small Strangers*, 121.

35. Gallup, "The Children's Museum as Educator," 372.

36. "Children's Museum League," *Children's Museum News*, November 1915.

37. "A Sheaf of Letters . . . ," *Children's Museum News*, May 1926, 273.

38. Hiner, "Seen but Not Heard"; West, *Kodak and the Lens of Nostalgia*.

39. Schlereth, *Cultural History and Material Culture*, 94–95.

40. Kohlstedt, *Teaching Children Science*, 178; Kramer, *Maria Montessori*, 15.

41. Higonnet, "A Golden Age," chap. 3 in *Pictures of Innocence*, 50–71.

42. Ibid.

43. Fass, *The Damned and the Beautiful*.

44. Stearns, "Obedience and Emotion."

45. "Do Children Appreciate the Lectures?," *Children's Museum News*, December 1913, 18.

46. See, for example, Lewis Hine, "Making Human Junk," *Child Labor Bulletin* 3 (1914–15): 37.

47. Dimock, "Priceless Children," 7–22.

48. "An Americanization Program," *Children's Museum News*, February 1920; "Americanization Lessons," *Children's Museum News*, March 1920.

49. Klapper, *Small Strangers*, 133.

50. "Museum Tests," *Children's Museum News*, April 1922, 52–53.

51. "Library Notes," *Children's Museum News*, February 1917.

52. "Bird Walks to Prospect Park," *Children's Museum News*, December 1913, 22.

53. "Children's Museum League."

54. "The Children's Museum," *Junior Eagle*, July 19, 1914.

55. "Prize Contests," n.d., BCM.

56. Gallup, "A Children's Museum."

57. "Lectures," *Children's Museum News*, May 1915.

58. "Working for Prizes," *Children's Museum News*, February 1917, 103.

59. "A Series of Field Trips," *Children's Museum News*, January 1919.

60. The mitigation of children's consumptive impulses through instruction is a major theme in Lisa Jacobson's work and one that I discuss in Chapter 2 of this book. Jacobson, *Raising Consumers*.

61. C. R., "One Phase of the Summer's Work," *Children's Museum News*, November 1917, 10. See also discussions about greed and covetousness in "A Woodcraft Hike," *Children's Museum News*, May 1922, 59, and "Field Trips for 1922," *Children's Museum News*, 1922, 5.

62. "Days A-Field and What They Mean," *Children's Museum News*, November 1920, 2.

63. "Summer Work in the Busy Bee Room," *Children's Museum News*, October 1916.

64. "Mineral Study in the Museum," *Children's Museum News*, November 1917.

65. "A Good Way to Spend the Summer," *Children's Museum News*, January 1919.

66. "Library Notes," *Children's Museum News*, February 1917.

67. "Requirements for Certificate in Bird Study," *Children's Museum News*, March 1918, 22.

68. "The Children's Museum League," *Children's Museum News*, May 1915.

69. "Children's Museum League," *Children's Museum News*, March 1920.

70. For thrift's importance to middle-class American values in the 1910s and 1920s, see Yarrow, *Thrift*.

71. Aaron Alcorn writes about this dynamic in his work on model airplanes and boyhood culture in the interwar years, pointing to the gap between adult expectations that aeromodeling be "a site of social reproduction" and the actuality of boys' practice: "Adults who gave life to their own particular conceptions of boyhood . . . discovered that boys often had differing views of their own." Alcorn, "Flying into Modernity."

72. Reid, "The Children's Wonder-House," 30.

73. "The Children's Museum," *Junior Eagle*, January 18, 1914.

74. A similar pattern of conditional inclusion can be seen later in the museum's his-

tory: in the 1930s the Brooklyn Children's Museum's Brooklyn Pick and Hammer Club, made up of patrons interested in mineralogy, included several young women, whose presence is recorded through their contribution to the newsletter that the club produced (*Pay Dirt*). The young women, however, appear not to have been invited on the field trips that the club's members lovingly recounted in the same newsletter, which served as important sites of community formation. I write about this club and its newsletter in the article "Writing a 'Wonderland' of Science: Child-Authored Periodicals at the Brooklyn Children's Museum, 1936–1946."

75. Douglas, *Inventing American Broadcasting*. For more on youth participation in wireless culture, see Michele Hilmes, *Radio Voices*, especially chap. 2, "How Far Can You Hear?"

76. Kohlstedt, *Teaching Children Science*, 119–20.

77. "New Books in the Library," *Children's Museum News*, December 1933.

78. Gallup, "The Children's Museum as Educator," 375.

79. Reid, "The Children's Wonder-House," 34.

80. Ibid., 30. Another article from an unnamed paper that has been preserved in the Brooklyn Children's Museum's archives also mentioned a museum patron, fourteen-year-old Andrew Bostwick, who fixed whatever "bell or electric light" might be out of order, making it so that Anna Gallup "rarely need[ed] to call an electrician." See "The Children's Museum," January 25, 1914.

81. "Wireless," *Children's Museum News*, December 1917.

82. Anna Billings Gallup, "A Mother's Account of a Children's Museum Lad," *Children's Museum News*, January 1914, 26–28.

83. "The Wireless Station," *Children's Museum News*, January 1915.

84. "Brooklyn Boys Radio Chiefs," *Brooklyn Daily Times*, October 25, 1915, BCM.

85. Reid, "The Children's Wonder-House."

86. "The Wireless Station."

87. "The Return of Our Boys," *Children's Museum News*, May 1919, 62; "A Promotion," *Children's Museum News*, December 1918, 16.

88. "The Wireless Class," *Children's Museum News*, October 1918, 5.

89. "In a Lively Sector," *Children's Museum News*, November 1918, 15. For more on American ideologies of technological superiority and the assumed accompaniment of cultural dominance, see Adas, *Dominance by Design*.

90. Gallup, "A Children's Museum and How Any Town Can Get One."

91. "Museum 'Alumni,'" *Children's Museum News*, November 1929.

Chapter 2

1. Frayling, *Mad, Bad, and Dangerous?*, 146; "Billy Bookworm," *Boys' Life*, June 1916, 40.

2. Wise, *Young Edison*, 22.

3. Lienhard, *Inventing Modern*, 191–92.

4. Keene, *Science in Wonderland*; Kohlstedt, "Parlors, Primers, and Public Schooling"; Pandora, "The Children's Republic of Science."

5. Bowler, *Science for All*; LaFollette, *Making Science Our Own*; LaFollette, *Science on the Air*; Hansen, *Picturing Medical Progress from Pasteur to Polio*; Shapin, *The Scientific Life*.

6. On contestations of scientism in the public sphere, see Colgrove, "'Science in a Democracy'"; Marsden, *Fundamentalism and American Culture*; Lederer, *Subjected to Science*; and Slotten, "Humane Chemistry or Scientific Barbarism?" On the growing ideological power of science, see Tobey, *The American Ideology of National Science*.

7. Rae, "Application of Science to Industry."

8. Hansen, *Picturing Medical Progress from Pasteur to Polio*, 128.

9. Kuznick, "Losing the World of Tomorrow"; Rydell, *World of Fairs*; Terzian, "The 1939–1940 New York World's Fair."

10. Luey, *Expanding the American Mind*; Rubin, *The Making of Middlebrow Culture*.

11. For more on the history of Science Service, founded in 1920, see Chapter 3.

12. "The Francis Bacon Award," n.d., Box 41, Folder 11, SA.

13. Quoted in Kramer, *Maria Montessori*, 131–32.

14. Op de Beeck, *Suspended Animation*, 13–14.

15. Maude Dutton Lynch, "Books Children Like and Why . . . ," *Parents*, November 1930, 23.

16. Rose, *The Case of Peter Pan*, 9.

17. Tebbel, "For Children, with Love and Profit," 16.

18. Mitchell, *Here and Now Story Book*, 25.

19. Ibid., 28.

20. Marcus, *Minders of Make-Believe*, 101–3.

21. Cited in ibid., 121.

22. Mitchell, *Here and Now Story Book*, 31.

23. Grolier, the book's American publisher, claimed in 1946 that they had sold more than 3,500,000 sets of the book since the first edition. Hammerton, *Child of Wonder*, 130. For the bedtime anecdote, see Antler, *Lucy Sprague Mitchell*, 265.

24. Rudolph, "Turning Science to Account"; Heffron, "The Knowledge Most Worth Having."

25. Op de Beeck, *Suspended Animation*, 10.

26. Douglas, *Inventing American Broadcasting*; Haring, *Ham Radio's Technical Culture*; Oldenziel, "Boys and Their Toys"; Pursell, "Toys, Technology, and Sex Roles in America."

27. On model airplanes and consumption, see Alcorn, "Flying into Modernity."

28. Smith, "The Evolution of the Chemical Industry," 141–50.

29. Rhees, "The Chemists' Crusade"; Slotten, "Humane Chemistry or Scientific Barbarism?"

30. Slosson, *Creative Chemistry*; LaFollette, *Making Science Our Own*, 85.

31. "To Make Money: Use Chemistry," *Literary Digest*, October 29, 1927.

32. Al-Gailani, "Magic, Science and Masculinity," 378.

33. See Cook, *The Commodification of Childhood*; Jacobson, *Raising Consumers*.

34. Marchand, *Advertising the American Dream*, 298–99.

35. Ruth L. Frankel, "Choosing the Right Toys," *Hygeia*, December 1931, 1407. Frankel attended a meeting of the Women's International League for Peace and Freedom, at which a display table held scientific toys. "It can easily be seen that the table full of scientific toys and objects was more than a table full of valuable apparatus," she wrote. "No child, interested enough to attempt to orient himself in a timeless, measureless universe, could ever again be completely arrogant or completely convinced that the universe revolved about him. Likewise, the child who is given one of the fascinating new biology sets or a chemistry outfit is lifted far from a local point of view. Science is universal. Scientists pride themselves on working for humanity and rejoice in bestowing their discoveries on the world; so the child whose experiments lead him to the many fields of scientific adventure is likely to step beyond boundaries and apart from patriotic segregation."

36. Cross, *Kids' Stuff*, 60.

37. "Inspiration," *Playthings*, December 1916.

38. Thomas K. Black, "The Romance of Toys," *Playthings*, June 1917, 86.

39. Cavallo, *Muscles and Morals*; Jackson, "Fit Body, Fit Mind"; Montoya, "Model Schools and Field Days"; Paris, *Children's Nature*; Macleod, *Building Character in the American Boy*; Valentine, "Playing Progressively?"

40. A. C. Gilbert Company, "Sacrifice! Sacrifice to the Bone for Ourselves, but Don't Sacrifice the Children's Christmas!" *Playthings*, August 1918.

41. A. C. Gilbert Company, "Enter the Discriminating Shopper," *Playthings*, November 1920.

42. "Every Toy Salesman Should Know," *Playthings*, November 1919; Lester G. Herbert, "Know and Sell—The Right Toy for Each Child," *Playthings*, June 1922.

43. See Cook, *The Commodification of Childhood*, 11. Developmentalism, Cook argues, "has served as the quintessential mode for the adjudication of what may be legitimized as beneficial to children"; starting in the 1920s and 1930s, the popularity of this mode of understanding childhood, which "posits predictable movement through specifiable and sequential stages of the early life course," meant that mothers were relegated to a middle position between children and the market, "softening the blow of commerce on the social and moral value of their children" by identifying the correct objects for children to consume.

44. "Teaching Children," *Playthings*, January 1919, 198.

45. Ethel Kawin, *Wise Choice of Toys*), 49–50.

46. Hannibal Coons, "Toy Tycoon," *Advertising & Selling*, November 1946.

47. Lionel Corporation, "Sweeping the Country," *Playthings*, September 1941.

48. A. C. Gilbert Company, "The 'High' Cut Out of Prices," *Playthings*, December 1920.

49. Keene, "'Every Boy and Girl a Scientist.'"

50. Gilbert, *The Man Who Lives in Paradise*, 136. Chemcraft, on the other hand, advertised in schoolbooks. "Chemcraft and Other Porter Products Sell at All Seasons," *Playthings*, February 1921.

51. Al Block to Peter Weinberg, March 23, 1967, A. C. Gilbert Papers (MS 1618), Manuscripts and Archives, Yale University Library.

52. Gilbert, *The Man Who Lives in Paradise*, 270–71.

53. John Anderson, "A Tale of Tricks, Toys and Tracks," *Merchandising News*, March 1952. Gilbert bought the American Flyer train line in 1938.

54. *Gilbert Toys Presents . . . Science Leads the Way!* (New Haven, Conn.: A. C. Gilbert Company, 1959).

55. A. C. Gilbert Company, "Over the Top for 1918," *Playthings*, n.d.

56. Treat Johnson, Elbert M. Shelton, and A. C Gilbert, *Gilbert Chemistry for Boys* (A. C. Gilbert Company, New Haven, Conn., 1936).

57. A. C. Gilbert Company, "Revealing the New Mystifying Secrets of Chemistry," *Boys' Life*, December 1930.

58. Johnson, Shelton, and Gilbert, "Gilbert Chemistry for Boys," 24.

59. Porter Chemical Company, *How to Be a Boy Chemist* (Porter Chemical Company, Hagerstown, Md., 1923), 10.

60. "Microset Instructions for Model No. 2MX" (Carolyn Manufacturing Co., Inc., New York, N.Y., n.d.), 1, 9, Object Collections, 2006.069.001, CHF.

61. *It's Fun to Be a Boy Engineer . . . a Boy Chemist . . . a Boy Scientist . . . a Boy Magician* (A. C. Gilbert Company, New Haven, Conn., 1940), 2.

62. "Chemcraft Experiment Book No. 2" (Porter Chemical Company, Hagerstown, MD, n.d.), 2, Object Collections, GB98:09.003, CHF.

63. *Youth Looks at Science and War* (Washington, D.C.: Science Service and Penguin, 1942); Hickam, *Rocket Boys*, 140.

64. "Lionel-Porter Glass Blowing Manual" (Lionel Corporation, Hagerstown, MD, 1937), 15, 28, Object Collections, 2005.079, CHF.

65. Porter Chemical Company, *Home Experiments in Science and Magic* (Porter Chemical Company, Hagerstown, Md., 1937), 16.

66. "Club News," *Chemcraft Science Magazine*, July 1939, 4–5, Object Collections, GB98:09.054, Chemistry Set Paper Ephemera, CHF.

67. "Club News," *Chemcraft Science Magazine*, December 1939, 20, ibid.

68. James L. Waters, interview by Arnold Thackray and Arthur Daemmrich, August 21, 2002, 7, Oral History Transcript No. 0262, CHF.

69. Lienhard, *Inventing Modern*, 191–92.

70. Dillon Roberts, "He Supplies Santa Claus," *Saga Magazine*, December 1950, 83.

71. "Chemcraft Experiment Book No. 2," 1.

72. Ibid., 2.

73. Morris Meister, founder of the Bronx High School of Science and booster of science clubs and fairs in New York City in the interwar years, took cues from the early success of clubs convened by Meccano, Gilbert, and Chemcraft, looking to their thriving networks as a model. Terzian, *Science Education and Citizenship*, 11.

74. "Club News," *Chemcraft Science Magazine*, June 1940, 4, Object Collections, GB98:09.054, Chemistry Set Paper Ephemera, CHF.

75. "The Chemcraft Science Club," *Chemcraft Science Magazine*, December 1937, ibid.

76. Porter Chemical Company, "Chemcraft Announcement," *Playthings*, January 1917. As far as I can tell, the use of the construction "boys and girls" in Chemcraft's advertising to toy buyers was intended to indicate the broadness of Chemcraft's fan base; the clubs writing into the Chemcraft magazine never seemed to mention girls as members.

77. Porter Chemical Company, "Harry K. Phillips Has a Message for Every Toy Buyer," *Playthings*, March 1922.

78. *It's Fun to Be a Boy Engineer . . . a Boy Chemist . . . a Boy Scientist . . . a Boy Magician*, 3.

79. Al-Gailani, "Magic, Science and Masculinity."

80. Jacobson, *Raising Consumers*, 119; Matt, "Children's Envy."

81. Watson, *The Man Who Changed How Boys and Toys Were Made*, 129.

82. *It's Fun To Be A Boy Engineer*, 14.

83. "Chemcraft Chemist and Handbook of Chemistry" (Porter Chemical Company, Hagerstown, Md.), 1, Object Collections, GB98:09.054, Chemistry Set Paper Ephemera, CHF.

84. Ibid., 4.

85. "Club News," *Chemcraft Science Magazine*, July 1939, 5.

86. "The Chemcraft Science Magazine: Official Organ of the Chemcraft Science Club, Vol. 48, No. 1" (Porter Chemical Company, Hagerstown, MD, n.d.), Object Collections, 2005.043, CHF.

87. Ibid., 2.

88. "The Chemcraft Science Club," 5. Not all young scientists earned their equipment through honest enterprise. The parents of Philip Eaton (b. 1936), later a professor of chemistry at the University of Chicago, realized this when young Philip, infatuated with his basement lab, started "borrowing" money from his father on the sly to order equipment that he wanted. This resulted in a year-long lab shutdown and what Eaton later termed "a valuable lesson in honesty—even if I was pursuing thievery in a higher interest." Philip E. Eaton, interview by James G. Traynham, January 22, 1997, 2, Oral History Transcript No. 0152, CHF.

89. A. C. Gilbert Company, "Gilbert Boy Chemists Lead in Thrilling Scientific Discoveries," *Boys' Life*, December 1935.

90. Morus, "Worlds of Wonder"; Nadis, *Wonder Shows*.

91. "Chemical Tricks, for Home Amusement and Instruction," published by F. Tousey, New York City, 1894, cited in Al-Gailani, "Magic, Science and Masculinity," 374.

92. Perkins, *The Reform of Time*.

93. "Chemcraft Chemical Magic" (Porter Chemical Company, Hagerstown, MD, 1937), 2, Object Collections, GB98:09.003, CHF.

94. Porter Chemical Company, *How to Be a Boy Chemist*, 14.

95. "Chemcraft Chemical Magic," 6.

96. Ibid., 7.

97. Gorton Adams, "Magic and More Magic," *4-H Horizons*, April 1939.

98. "Chemcraft Chemical Magic," 19.

99. Ibid., 22.

100. Ibid., 25.

101. Ibid., 19.

102. Ibid., 24.

103. By 1956, a Chemcraft manual that included a shortened version of the "chemical magic" instructions omitted the section on the "slave." Harold M. Porter, "Porter Chemcraft Junior Experiment Manual" (Porter Chemical Company, Hagerstown, MD, 1956), Object Collections, GB98:09.004, CHF.

104. *Chemcraft Chemical Magic: Mystifying Magical Demonstrations* (Hagerstown, Md.: Porter Chemical Company, 1937), 119.

105. Rossiter, *Women Scientists in America*, 119–20.

106. Tolley, *The Science Education of American Girls*. See chap. 8, "Physics for Boys."

107. "Proposing, a New Field for Toys—Sell Toys to the Girl!" *Playthings*, June 1921.

108. Coons, "Toy Tycoon."

109. "Chemcraft and Other Porter Products Sell at All Seasons," *Playthings*, February 1921.

110. "The New Products of the Porter Chemical Company," *Playthings*, July 1920.

111. "Merchandise, Markets, and Men," *Playthings*, June 1917.

112. Petersham and Petersham, *The Story Book of Gold*.

113. Pryor and Pryor, *The Steel Book*, 26.

114. Pryor and Pryor, *The Streamline Train Book*, 22–26.

115. Tolley, *The Science Education of American Girls*, 200.

116. *Chemcraft, the Chemical Outfit: No. 5 Experiment Book* (Hagerstown, Md.: Porter Chemical, 1928).

117. Foy and Schlereth, *American Home Life*, 75.

118. Ibid., 88.

119. Kristen Haring's study of ham radio culture does attend to the way that families allocated spaces within the home for radio hobbyists both young and old, though she focuses on the midcentury period; I have been unable to find work on the history of the in-home children's chemistry lab in the United States. Haring, *Ham Radio's Technical Culture*.

120. See chap. 2, "Bad Boys and Men of Culture," in Kidd, *Making American Boys*.

121. Feynman, *"Surely You're Joking, Mr. Feynman,"* 16–17; Sacks, *Uncle Tungsten*, 273–74.

122. Burnham, "Development of Environmentalism," 60.

123. Tarr and Tebeau, "Housewives as Home Safety Managers," 220.

124. Burnham, "Why Did the Infants and Toddlers Die?," 819.

125. See chap. 3, "Modernizing the House and Family," in Clark, *The American Family Home*.

126. Tarr and Tebeau, "Managing Danger in the Home Environment" 797.

127. Innes, "Men Who Made the Railways," 611; Atkinson, *The How and Why Library*,

frontispiece; *The New Wonder World*, frontispiece; Hartman, *The World We Live in and How It Came to Be*, 241.

128. Hartman, *The World We Live in and How It Came to Be*, 240.

129. Kohlstedt, "Parlors, Primers, and Public Schooling," 430.

130. Angier, *The Canon*, 1–3.

131. Patterson, *Robert A. Heinlein*, 28.

Chapter 3

1. *Life*'s package joined a spate of articles analyzing and extolling the Soviet educational system in professional and popular-interest publications in 1957 and 1958. Lawrence G. Derthick, "The Russian Race for Knowledge," *School Life*, June 1958.; Holland, "The Current Challenge of Soviet Education," *School and Society* 86, no. 2134 (June 7, 1958): 261–63; Howland Sargeant, "Soviet Challenge in Education," *Commonweal*, April 26, 1957; Harry Schwartz, "Soviet Gains in Science: A View That Incentive for Technical Work Is Greater in Russia Than in U.S. Educational Differences," *New York Times*, October 13, 1957; "Education in the USSR," *School Life*, December 1957.

2. "Schoolboys Point Up a U.S. Weakness," *Life*, March 24, 1958.

3. "The Waste of Fine Minds," *Life*, April 7, 1958, 91.

4. "Schoolboys Point Up a U.S. Weakness"; "Crisis in Education," *Life*, March 24, 1958; "The Waste of Fine Minds."

5. Sloan Wilson, "It's Time to Close Our Carnival," *Life*, March 24, 1958, 37.

6. Despite intense rhetoric like Wilson's, public responses to *Life*'s panicked articles on the crisis of education were split. Some response letters from readers agreed with the alarmism, but others wrote, "Count the smiles of the Russians and those of the Americans. Surely this must be worth something" (Mildred Jensen); "As long as Steve has a chick like Penny and Alexei has homework, I figure that Steve is the winner" (Keith Echlaw). "Letters to the Editors," *Life*, April 14, 1958.

7. Rudolph, *Scientists in the Classroom*; Hartman, *Education and the Cold War*; Clowse, *Brainpower for the Cold War*.

8. As Lynn Spigel puts it, concerns about childhood in the postwar era were "typically articulated in terms of psychological discourses of personality development in relation to larger national questions about authoritarianism and freedom." Spigel, *Welcome to the Dreamhouse*, 224.

9. See Lecklider, *Inventing the Egghead*.

10. Mead and Metraux, "Image of the Scientist"; "Where Do Our Young People Get Such Crazy Ideas about What Scientists Do?" *Saturday Evening Post*, November 30, 1957.

11. David C. Beardslee and Donald D. O'Dowd, "The College-Student Image of the Scientist," *Science*, n.s., 133, no. 3457 (March 31, 1961): 997. The gap between perceptions of "intellectual" scientists and "intelligent" engineers (in Richard Hofstadter's terms) can be seen in the students' evaluations of engineers, who got higher scores on the descriptors "clean cut," "plays poker," and "has good taste." Additionally, the engineer was "believed more likely to have a pretty wife" (998). W. W. Hagerty, "Students' Images

of Scientists and Engineers," *BioScience* 14, no. 3 (March 1, 1964): 20–23, another study on college students' rationales for leaving science, concluded, "Experiences that drive them out have far less to do with science as either a method or an epistemological system than they do with science as a social institution." Hagerty's study is discussed in Edgar Z. Friedenberg, "Why Students Leave Science," *Commentary* 3 (August 1961): 144–55.

12. Watson Davis, "Science Clubs and the Future" (Science Teachers of Cleveland, Ohio, October 25, 1948), 3–4, SA.

13. Medovoi, *Rebels*, 23.

14. Robert MacCurdy, "Characteristics and Backgrounds of Superior Science Students," *School Review*, 1956, 67.

15. Cole, *Encouraging Scientific Talent*, 11.

16. Wang, *In Sputnik's Shadow*.

17. Leslie, *The Cold War and American Science*, 1.

18. Shapley quoted in Kevles, *The Physicists*, 355; Wiener quoted in Shapin, *The Scientific Life*, 82.

19. Wang, *American Science in an Age of Anxiety*.

20. Kaiser, "The Postwar Suburbanization of American Physics," 862; Shapin, *The Scientific Life*, 80–83.

21. Kuznick, "Losing the World of Tomorrow."

22. Burnham, *How Superstition Won and Science Lost*.

23. See, for example, the memoir of West Virginia resident Homer H. Hickam, who saw Wernher von Braun interviewed on television and became increasingly interested in rocketry as a result. Hickam, *Rocket Boys*.

24. Shapin, *The Scientific Life*, 78.

25. LaFollette, *Making Science Our Own*, 128–29; 138–39.

26. Lienhard, *Inventing Modern*, 248.

27. Hofstadter, *Anti-Intellectualism in American Life*, 25–26.

28. Shapin, *The Scientific Life*, 173.

29. Bush, *Science, the Endless Frontier*, 18–19.

30. Wilson quoted in Daniel S. Greenberg, *The Politics of Pure Science* (New York: New American Library, 1968), 273, cited in Kevles, *The Physicists*, 383.

31. Shapin, *The Scientific Life*, 101.

32. In addition to Davis, committee members included James B. Conant of Harvard; Robert E. Doherty, president, Carnegie Institute of Technology; and Harlow Shapley of the Harvard College Observatory (among others). Bush, *Science, the Endless Frontier*. See p. 136 for a complete list of committee members.

33. Ibid., 159.

34. Terzian, *Science Education and Citizenship*, 97.

35. Spring, *The Sorting Machine*, 83–86.

36. Cremin, *The Transformation of the School*; Hartman, *Education and the Cold War*.

37. Rickover, *Education and Freedom*.

38. Rudolph, *Scientists in the Classroom*.

39. LaFollette, *Science on the Air*, 46–47; W. E. Ritter, "Science Service as One Expression of E.W. Scripps's Philosophy of Life," *Science News-Letter* 10, no. 298 (1926); Scripps, *Damned Old Crank*.

40. Science Service drew its trustees from the National Academy of Science, the National Research Council, the American Association for the Advancement of Science, the E.W. Scripps estate, and the journalistic profession; the final count of the governing board was ten scientists and five journalists, a makeup intended to signal to scientists that Science Service deserved their respect. Davis, "The Rise of Science Understanding," *Science*, September 3, 1948; LaFollette, *Science on the Air*, 59; Edwin E. Slosson, "A New Agency for the Popularization of Science," *Science*, n.s., 53, no. 1371 (April 8, 1921): 322.

41. LaFollette, *Science on the Air*, 55. On Slosson's philosophy of popular science, see David Rhees, "A New Voice for Science," in particular pt. 3, "The Popular Science of Edwin E. Slosson."

42. Davis, "The Rise of Science Understanding," 241. By 1948, circulation was "a couple of hundred newspapers and other publications with a readership of about 10,000,000."

43. LaFollette, *Science on the Air*, 61. In 1948, Davis reported that the *Science News-Letter* had a circulation of 50,000. Davis, "The Rise of Science Understanding," 243.

44. See, for a very few examples, Watson Davis, "Critical Shortage," *Science News-Letter* 48, no. 8 (August 25, 1945); Watson Davis, "The Frontiers of Science Are Still Endless" (Commencement, Bronx High School of Science, Bronx, NY, June 25, 1946), SA; Watson Davis, "Science and Libraries Today" (New Jersey Library Association, Buck Hill Falls, PA, March 20, 1947), SA; Watson Davis, "Science, Education, and the Press" (Southern Interscholastic Press Association, Washington, D.C., April 25, 1947), SA.

45. Lewenstein, "The Meaning of 'Public Understanding of Science.'"

46. Davis, "The Rise of Science Understanding," 241.

47. Ibid., 244; Watson Davis to Pierre Auger, October 25, 1951, Box 305, Folder 1, SA; "UNESCO International Meeting of Science Club Leaders," July 15, 1949, Box 446, Folder 33, SA. The finding aid for the Smithsonian's Science Service records, authored by Marcel LaFollette, mentions that Davis was also a proponent of the dissemination of Interlingua, an international scientific language (Record Unit 7091, Science Service, Records, 1902–1965, SA).

48. When Science Service took over the American Institute's Science Clubs of America in 1941, there were 700 clubs nationwide; by 1945, there were 7,419. Terzian, *Science Education and Citizenship*, 82–87.

49. Davis, "Science Clubs and the Future," 3–4.

50. Davis, "The Rise of Science Understanding," 6.

51. Davis, "Science Clubs and the Future," 4.

52. Davis, "Science Teaching and Science Clubs Now and Postwar," *School Science and Mathematics*, March 1945, 257–64.

53. "Science Service Aids to Youth," n.d., Box 401, Folder 35, SA.

54. Davis, "Science Teaching and Science Clubs Now and Postwar," 260; Terzian, *Sci-*

ence Education and Citizenship, 83. Davis wrote in this 1944 speech that Science Clubs of America contained 6,000 clubs, with about 150,000 individual members.

55. Science Service, "Encouraging Science Talent . . . The National Science Youth Program," 1958, Box 401, Folder 35, SA.

56. "Aiding Young Scientists," *Science News-Letter* 54, no. 16 (October 16, 1948): 246.

57. Berger, *The Young Scientists*, 17.

58. Phares and Westinghouse Electric Corporation, *Seeking—and Finding—Science Talent*, 3; Stanley Edgar Hyman and St. Clair McKelway, "The Time Capsule," *New Yorker*, December 5, 1952.

59. The fair was organized by the American Institute of the City of New York, and the student exhibits were developed through the Science and Engineering Clubs. Terzian, "The 1939–1940 New York World's Fair." In other Westinghouse projects, the Westinghouse Educational Foundation sponsored a summer institute program for teachers at MIT in 1949, the George Westinghouse Science Writing Awards (cosponsored with the American Association for the Advancement of Science) starting in 1945 (Rudolph, *Scientists in the Classroom*, 61), and a Saturday enrichment program called the Westinghouse Science Honors Institute, beginning in 1958 ("Saturday's Children at Westinghouse Labs," *Business Week*, May 1960, 104–6).

60. Sevan Terzian argues that postwar American science education was the site of a clash of rationales: "Science education for democratic citizenship and informed evaluation of consumer products . . . a practical curriculum that applied theoretical principles to aspects of daily living" versus "rigorous, discipline-based courses with the brightest students who had been carefully selected on the basis of their academic achievements and intellectual promise." Terzian, "'Adventures in Science,'" 310–11.

61. Berger, *The Young Scientists*, 18.

62. See mentions of contestants with this history in Science Service, "Supplementary Information for the Seventh Annual Science Talent Search," January 22, 1948, Box 394, Folder 7, SA; Science Service, "Supplementary Information for the Twelfth Annual Science Talent Search," 1953, Box 401, Folder 5, SA; Science Service, "Supplementary Information for the Seventeenth Annual Science Talent Search," February 1958, Box 401, Folder 39, SA; Science Service, "Young Scientists Series," February 6, 1958, Box 401, Folder 39, SA; Science Service, "Science Service Citations Go to Seventeen Students," n.d., Box 401, Folder 39, SA.

63. Davis, "The Interpretation of Science through Press, Schools, and Radio," *Engineers' Bulletin*, February 1952.

64. Margaret E. Patterson, "Presentation of Science Talent Search Plaque to Miss Marion Cecile Joswick" (Brooklyn, N.Y., n.d.), 1, Box 289, Folder 3, SA.

65. Davis, "Science Clubs and the Future," 3–4.

66. Davis, "The Next Half Century's Human Talents Will Be Assayed to Fit People to Best Life Work," *Science Page*, February 29, 1950. 1.

67. Harold A. Edgerton, "Again the *Searchlight* Swings Your Way," *Science Talent Searchlight*, February 1948, Box 446, Folder 20, SA.

68. *Youth Looks at Science and War* (Washington, D.C.: Science Service and Penguin, 1942), iv.

69. The company explicitly stated (in 1944, upon founding its philanthropic arm, the Westinghouse Educational Foundation) that it hoped to curry favor among college students who would develop warm feelings toward the provider of scholarships and then be inclined to deal well with the company when they moved into positions of power in the business world. Terzian, *Science Education and Citizenship*, 89.

70. Herman, *The Romance of American Psychology*.

71. Cohen-Cole, *The Open Mind*, 1.

72. Ball, "The Politics of Social Science in Postwar America," 82.

73. Solovey, "Cold War Social Science," 1–2.

74. Bycroft, "Psychology, Psychologists, and the Creativity Movement," 199.

75. Cohen-Cole, "The Reflexivity of Cognitive Science," 126.

76. Jenkins, "Dennis the Menace," 119.

77. See psychoanalyst Martha Wolfenstein's contemporaneous analysis of the trend away from recommendations of discipline and toward the promotion of "fun morality" over several decades of the Infant Care bulletin of the United States Children's Bureau. Wolfenstein, "The Emergence of Fun Morality" *Journal of Social Issues* 7, no. 4 (1951): 15–25. On the strong influences of ideals of creativity on postwar children's culture, see Ogata, "Creative Playthings"; Ogata, "Building Imagination in Postwar American Children's Rooms"; Ogata, *Designing the Creative Child*.

78. Bycroft, "Psychology, Psychologists, and the Creativity Movement," 199.

79. Calvin W. Taylor, "The 1955 and 1957 Research Conferences: The Identification of Creative Scientific Talent," *American Psychologist* 14 (1959): 100–102.

80. Specifically, the Rorschach Method and the Thematic Apperception Test. Roe, *The Making of a Scientist*, 7.

81. Brandwein, *The Gifted Student as Future Scientist*; Cole, *Encouraging Scientific Talent*; Knapp and Wesleyan University, *Origins of American Scientists*.

82. Kubie, "Some Unsolved Problems of the Scientific Career (Part 1)," *American Scientist* 41, no. 4 (October 1, 1953): 596–613; Kubie, "Some Unsolved Problems of the Scientific Caree (Part 2)," *American Scientist* 42, no. 1 (January 1, 1954): 104–12; Kubie, "The Fostering of Creative Scientific Productivity," *Daedalus* 91, no. 2 (April 1, 1962): 294–309.

83. Adorno, *Authoritarian Personality*.

84. Chorness, "An Interim Report on Creativity Research," 290.

85. Brandwein, *The Gifted Student as Future Scientist*, 52.

86. Ibid., 10–11.

87. Roe, "Personal Problems and Science," 134–36.

88. Getzels and Jackson, "The Highly Intelligent and the Highly Creative Adolescent," 172.

89. Roe, *The Making of a Scientist*, 92. Social scientists, however, were more active on the dating scene; Roe said of their interviews on the subject, "I have a decided impression, not only of much more dating, but also of much freer sexual activity generally."

90. Ibid., 235.

91. Roe, "Personal Problems and Science," 136.

92. Cattell, "The Personality and Motivations of the Researcher," 129–30.

93. Harold A. Edgerton, "Harold A. Edgerton, PhD: A Career in Industrial and Measurement Psychology," *Industrial & Organizational Psychology, Inc.*, http://www.siop.org /presidents/Edgerton.aspx (accessed June 20, 2012).

94. "Steuart Henderson Britt (1907–1979) Papers, 1927–1980." Finding aid at the Northwestern University Archives, Evanston, Ill.

95. Edgerton, "Harold A. Edgerton, PhD."

96. Harold A. Edgerton and Steuart Henderson Britt, "The First Annual Science Talent Search," *American Scientist* 31, no. 1 (January 1, 1943): 55–68; Harold A. Edgerton and Steuart Henderson Britt, "The Third Annual Science Talent Search," *Science* 99, no. 2573 (April 21, 1944): 319–20; Harold A. Edgerton and Steuart Henderson Britt, "Sex Differences in the Science Talent Test," *Science* 100, no. 2592 (September 1, 1944): 192–93; Harold A. Edgerton and Steuart Henderson Britt, "The Science Talent Search in Relation to Educational and Economic Indices," *School and Society* 63, no. 1628 (March 9, 1946): 172–75.; Harold A. Edgerton, Steuart Henderson Britt, and Ralph Norman, "Later Achievements of Male Contestants in the First Annual Science Talent Search," *American Scientist* 36, no. 3 (July 1, 1948): 403–14.

97. MacCurdy, "Characteristics of Superior Science Students."

98. Norman, "A Study of Scientifically Talented Boys."

99. "Personal Data Blank: The First Annual Science Talent Search" (Science Service, Inc., 1942), Box 40, Folder 3, PP.

100. Edgerton, "Harold A Edgerton, PhD." In 1942 and 1943, the first two years of the contest, a combined total of 6,656 students entered. Edgerton, "Science Talent: Its Early Identification and Later Development," *Journal of Experimental Education* 34, no. 3 (April 1, 1966): 90.

101. Harold A. Edgerton, "How Science Talent Winners Were Chosen Told by Judge," *Science News-Letter* 42, no. 4 (July 25, 1942): 55.

102. Brandwein, "The 'Science' Talent Search," *Science*, n.s., 101, no. 2614 (February 2, 1945): 117.

103. Edgerton, "How Science Talent Winners Were Chosen Told by Judge," 54.

104. Edgerton, "Harold A Edgerton, PhD."

105. "Test Your Science Ability with Sample Problems," *Science News-Letter* 55, no. 5 (January 29, 1949): 72.

106. For example, in the 1943 test, the paragraphs were about the following topics: ants as vectors of disease; the relative visibility of stars and planets; the dental formulas used to classify the teeth of vertebrates; the meaning of operationism in the scientific method; the theory of panspermia and the possibility of spontaneous generation of life; the nature of conditioned responses; the anatomical features of the skin; the life cycle of molds; Archimedes's principle; the action of cumulonimbus clouds in precipitation; the circulation system and the liver; and the anatomy of vertebrate embryos. "The Second

Annual Science Talent Search Science Aptitude Examination" (Science Service, 1943), Box 40, Folder 3, PP.

107. "Quiz for Science Ability," *Science Page*, January 29–February 4, 1950, 1.

108. Fletcher G. Watson, "Analysis of a Science Talent Search Examination," *Science Teacher*, November 1954, 274–76.

109. Frank Jewett to Watson Davis, March 6, 1947, Box 289, Folder 3, SA.

110. "Notes on Response to Jewett Re: Science Talent Search Tests," n.d., Box 289, Folder 3, SA.

111. Watson Davis to Frank Jewett, March 14, 1947, Box 289, Folder 3, SA.

112. Hoffmann, *The Tyranny of Testing*; Lemann, *The Big Test*, 99–101.

113. Hoffmann, "Some Remarks concerning the First Annual 'Science Talent Search,'" *American Scientist* 31, no. 3 (July 1, 1943): 255–65.

114. Edgerton, "How Science Talent Winners Were Chosen Told by Judge," 54.

115. Whyte, *The Organization Man*, 212.

116. "Personal Data Blank."

117. Hoffmann, "Some Remarks."

118. Harold A. Edgerton and Steuart Henderson Britt, "Further Remarks regarding the Science Talent Search." *American Scientist* 31, no. 3 (Summer 1943): 264.

119. In a 1966 article, for example, Edgerton correlated scores on the STS exam with current career paths. Edgerton, "Science Talent."

120. Patterson, "Presentation of Science Talent Search Plaque to Miss Marion Cecile Joswick."

121. Shapin, *The Scientific Life*, 77; Glenn T. Seaborg, "Making the Creative Scientist," *Science News-Letter* 79, no. 11 (March 18, 1961): 170–72; "Wanted: Young Scientists," *Science News-Letter* 76, no. 6 (August 8, 1959): 82.

122. Paul Cloke, "New STSers," *Science Talent Searchlight*, February 1948, 32, Box 446, Folder 20, SA.

123. "Program for the Eleventh Annual Science Talent Search," March 28, 1952, Box 399, Folder 10, SA.

124. Ibid.

125. Science Talent Institute, "President Greets Young Talent Search Winners Today," 1955, Box 330, Folder 5, SA.

126. Science Service, "Forty Top Young Scientists Arrive in Nation's Capital," February 23, 1955, Box 330, Folder 5, SA.

127. Science Talent Institute, "Teen-Age Scientists of Tomorrow Learn to Chart Scientific Course," March 2, 1956, Box 330, Folder 5, SA.

128. "Program for the Eleventh Annual Science Talent Search."

129. Eugene F. Haugh, "Dear STSers," *Science Talent Searchlight*, February 1948, 34, Box 446, Folder 20, SA.

130. Science Clubs of America, *Science Talent Searchlight*, February 1948, 53, Box 446, Folder 20, SA.

131. Karl T. Compton, "Young Scientists and War," *Science News-Letter* 45, no. 12

(March 18, 1944): 180. In another example, near the end of his speech to the STS finalists at the 1949 gala dinner, chairman of the Red Cross Basil O'Connor listed the names
of scientists who had made important discoveries at a young age. "This is how humanity
progresses, in every field of endeavor," he said. "The young mind re-explores the established principles and picks up on the flaws previously overlooked. Or starting off on a
tangent, in pursuit of some original idea that may have no obvious practical application
or intrinsic value, suddenly ends up with an amazing basic contribution to the art or the
science of better living." Basil O'Connor, "Science and Humanity," March 7, 1949, Box
313, Folder 6, SA.

132. Kaiser, "The Postwar Suburbanization of American Physics," 868.

133. Harlow Shapley, "Scholarship Winners Told Beware of Bumptious Vanity," *Science
News-Letter* 42, no. 4 (July 25, 1942): 52.

134. "Young Scientists at Work," *Chemistry*, March 1951, 2.

135. Terzian, "'Adventures in Science.'"

136. Science Talent Institute, "Teen-Age Scientists Offer Cures for Threat to U.S.
Technological Survival," March 5, 1956, Box 330, Folder 5, SA. As Venable won the third-
place prize that year, his plea may not have been completely disinterested.

137. "STS 1942 Finalists Enjoy a Swimming Party," 1942, Society for Science and the
Public, Washington, D.C., http://www.flickr.com/photos/societyforscience/4973735587
/in/set-72157624916749128/ (accessed November 2, 2012); "Girl Scouts Represented
in the STS 1945," 1945, Society for Science and the Public, Washington, D.C., http://
www.flickr.com/photos/societyforscience/4973755563/in/set-72157624916800220/
(accessed November 2, 2012); "STS 1946 Finalists Get to Sit in a Senate Committee
Meeting," 1946, Society for Science and the Public, Washington, D.C., http://www
.flickr.com/photos/societyforscience/4974378814/in/set-72157624916821580/ (accessed
November 2, 2012); "STS 1946 Finalists on the Capitol Subway to Meet with the Vice
President Accompanied by Senator Kilgore," 1946, Society for Science and the Public,
Washington, D.C., http://www.flickr.com/photos/societyforscience/4974380282/in/set
-72157624916821580/ (accessed November 2, 2012); "STS 1946 Finalists Get to Sit in
a Senate Committee Meeting," 1946, Society for Science and the Public, Washington,
D.C., http://www.flickr.com/photos/societyforscience/4974378814/in/set-721576249
16821580/ (accessed November 2, 2012); "STS 1947 Judge, Dr. H. Edgarton Interviews
Finalist Vera Dyson-Hudson (Demerec)," 1947, Society for Science and the Public,
Washington, D.C., http://www.flickr.com/photos/societyforscience/4974386616/in
/set-72157624916843630/ (accessed November 2, 2012); "STS 1947 Finalists Visit President Harry Truman in the Oval Office," 1947, Society for Science and the Public, Washington, D.C., https://www.flickr.com/photos/societyforscience/4974387946/ (accessed
November 2, 2012); "STS 1950 Dr. Harlow Shapley, President of Science Service Speaks
with Finalists," 1950, Society for Science and the Public, Washington, D.C., http://www
.flickr.com/photos/societyforscience/4974436442/in/set-72157624916974252/ (accessed
November 2, 2012); "STS 1955 Finalists with President Dwight D. Eisenhower at the
White House," 1955, Society for Science and the Public, Washington, D.C., http://www

.flickr.com/photos/societyforscience/4974487502/in/set-72157624792514565/ (accessed November 2, 2012); "STS 1957 Finalists Sharing a Milkshake," 1957, Society for Science and the Public, Washington, D.C., http://www.flickr.com/photos/societyforscience /4974514838/in/set-72157624917196884/ (accessed November 2, 2012); "STS 1960 Gayle Wilson (Edlund) and David Brown Playing in the Snow," 1960, Society for Science and the Public, Washington, D.C., http://www.flickr.com/photos/societyfor science/4975022344/in/set-72157624918537338/ (accessed November 2, 2012); "Boy Scouts Represented in the STS 1945," 1945, Society for Science and the Public, Washington, D.C., http://www.flickr.com/photos/societyforscience/4974371808/in/set -72157624916800220/ (accessed November 2, 2012).

138. Science Talent Institute, "Teen-Age Scientists of Tomorrow Learn to Chart Scientific Course."

139. Science Service, "Pennsylvania Boy, Wisconsin Girl, Colorado Boy Take Top Talent Search Honors," February 28, 1955, Box 330, Folder 5, SA. See also coverage of Science Service's "research" on contestants in *Science Digest*, whose writer began the piece: "The promising young scientist turns out not to be the eccentric that people may think him to be." "Teen-Aged Scientists Unlike Stereotypes," *Science Digest*, May 1958.

140. Waterman, "Introduction," xvii.

141. Watson Davis, "Memorandum on National Science Fair," November 1949, 1, Box 446, Folder 37, SA.

142. Science Talent Institute, "Maryland, Illinois, and Georgia Boys Take Top Talent Search Honors," March 5, 1956, Box 330, Folder 5, SA.

143. "Top Young U.S. Scientists Chosen," *Science News-Letter* 77, no. 12 (March 19, 1960): 182.

144. A comic book included in a 1950s Gilbert chemistry set offers science as an alternative competitive activity for male students. In the narrative, a team of boys from a small high school, tired of being beaten on the football field by a rival, decides to turn to science, mounting a successful challenge to the City High science team using their Gilbert sets to vividly demonstrate scientific principles. See *Gilbert Toys Presents . . . Science Leads the Way!*

145. "Biographical Notes on Science Talent Search Contestants," 1953, Box 401, Folder 5, SA.

146. James B. Gibson, "Greetings," *Science Talent Searchlight*, February 1948, 26, Box 446, Folder 20, SA.

147. Russell Johnson, "Dear STSers," *Science Talent Searchlight*, February 1948, 27, Box 446, Folder 20, SA.

148. Edgerton and Britt, "The Third Annual Science Talent Search," 320. Perhaps Britt and Edgerton felt the need to emphasize the anonymity of the process because some states (and some schools) were overrepresented among trip winners; in 1944, for example, Britt, Edgerton, and the Service's Helen Davis wrote that one-third of trip winners in the first three years of the contest had come from New York State. Harold A.

Edgerton, Steuart Henderson Britt, and Watson Davis, "Is Your State Discovering Its Science Talent?," *Science Education* 28, no. 4 (October 1944).

149. Science Service, "Science Experiments Most Interesting Hobby," June 23, 1945, Box 391, Folder 18, SA.

150. Science Service, "Young Scientists Series," February 18, 1957, Box 402, Folder 6, SA.

151. "Biographical Notes on Science Talent Search Contestants," 1953, Box 401, Folder 5, SA.

152. Shirley Moore, "Raising a Future Scientist," *Science News-Letter* 76, no. 9 (August 29, 1959): 135.

153. Shirley Moore, "Science Youths Start Younger," *Science News-Letter* 84, no. 11 (September 14, 1963): 170.

154. Shirley Moore, "Bringing Up a Scientist," *Science News-Letter* 77, no. 23 (June 4, 1960): 362.

155. Thomas J. Fleming, "A Talk with a Talent Scout," *Reader's Digest*, June 1961, 34.

156. Besser, *Growing Up with Science*, 175–77.

157. "Aiding Young Scientists," 245. The state and national searches ran concurrently, rather than the state-level searches serving as a first barrier to entry for the national contest.

158. Margaret Patterson, "State Science Talent Searches, 1948–1949," 1949, 4, Box 446, Folder 37, SA.

159. Laura Vanderkam, "Sister Julia Mary Deiters: Planting Seeds for Science Education," *Scientific American*, October 13, 2008.

160. Critical ratios: 1942: 14.4; 1943: 16.6; 1944: 18.2.

161. Edgerton and Britt, "Sex Differences in the Science Talent Test," 193.

162. "STS 1950 Annual Capitol 'Pointing to the Future' Photo," 1950, Society for Science and the Public, Washington, D.C., http://www.flickr.com/photos/societyforscience /4974433238/in/set-72157624916974252/ (accessed November 2, 2012); "STS 1952 Finalists Louise Hay (Schmir) and Paul Forman 'Looking to the Future,'" 1950, Society for Science and the Public, Washington, D.C., http://www.flickr.com/photos/societyfor science/4973834599/in/set-72157624792395597/ (accessed November 2, 2012); "STS 1954 Finalists, Marguerite Bloom (Burlant) and Robert Rodden 'Look to the Future,'" 1954, Society for Science and the Public, Washington, D.C., http://www.flickr.com /photos/societyforscience/4974481728/in/set-72157624917106158/ (accessed November 2, 2012); "STS 1955 Finalists Carol MacKay (Myers) and Stephen Webb 'Looking into the Future' at the Capitol," 1955, Society for Science and the Public, Washington, D.C., http://www.flickr.com/photos/societyforscience/4973873769/in/set-721576247 92514565/ (accessed November 2, 2012); "STS 1955 Finalist Carol Cartwright (Hawkins) Was Not Only a Gifted Young Scientist but Also an Accomplished Seamstress," 1955, Society for Science and the Public, Washington, D.C., http://www.flickr.com/photos /societyforscience/4974487632/in/set-72157624792514565/ accessed November 2, 2012); "STS 1956 Capitol Steps 'Pointing to the Future' Photo," 1956, Society for Science and the Public, Washington, D.C., http://www.flickr.com/photos/societyforscience

/4973884951/in/set-72157624792544821/ (accessed November 2, 2012); "STS 1957 Finalists Sharing a Milkshake," 1957, Society for Science and the Public, Washington, D.C.,
http://www.flickr.com/photos/societyforscience/4974514838/in/set-7215762491719
6884/ (accessed November 2, 2012); "STS 1957 Finalists 'Point to the Future,'" 1957,
Society for Science and the Public, Washington, D.C., http://www.flickr.com/photos
/societyforscience/4974513634/in/set-72157624917196884/ (accessed November 2,
2012); "STS 1959 Finalists Daniel Banberg Jr. and Marion Madrid (Davis) 'Look to the
Future,'" 1959, Society for Science and the Public, Washington, D.C., http://www.flickr
.com/photos/societyforscience/4974528812/in/set-72157624917242322/ (accessed
November 2, 2012); "STS 1960 Female Finalists Look on at the Hope Diamond Displayed at the Smithsonian Natural History Museum," 1960, Society for Science and the
Public, Washington, D.C., http://www.flickr.com/photos/societyforscience/497440
8237/in/set-72157624918537338/ (accessed November 2, 2012).

163. Rudolph, *Scientists in the Classroom*, 60.

164. Kaiser, "The Postwar Suburbanization of American Physics," 877.

165. Puaca, *Searching for Scientific Womanpower*, 103–7.

166. Science Service, "Pennsylvania Boy, Wisconsin Girl, Colorado Boy Take Top
Talent Search Honors."

167. "Top Young U.S. Scientists Chosen," 182.

168. Science Service, "Supplementary Information for the Seventeenth Annual Science Talent Search."

169. Science Service, "Supplementary Information for the Twelfth Annual Science
Talent Search."

170. Science Service, "Supplementary Information for the Sixteenth Annual Nationwide Science Talent Search," February 6, 1957, Box 402, Folder 6, SA; Science Service,
"Young Scientists Series," February 18, 1957.

171. Edgerton, Britt, and Norman, "Later Achievements of Male Contestants," 404.

172. Edgerton, "Science Talent," 95–96.

173. Fleming, "A Talk with a Talent Scout," 34.

174. Elizabeth J. Foster, "Dear Fellow STS'ers," *Science Talent Searchlight*, February
1948, Box 446, Folder 20, SA.

175. Virginia March Kline, "Hi Gang!" *Science Talent Searchlight*, February 1948, Box
446, Folder 20, SA.

176. Science Clubs of America, "Science Talent Searchlight," 47.

177. Hersey, *The Child Buyer*. In a 1961 review of the book in *Science*, Margaret Mead
wrote: "This book . . . brings into sharp and shocking conjunction two salient aspects of
our present attitudes: our willingness to treat people as things — 'the human component
in systems design' — and our interest in the gifted child — whom otherwise a democratic
society would prefer to ignore — as a defense need." Mead, "Review," *Science*, February
24, 1961, 573.

178. Schofield-Bodt, "A History of Children's Museums," 5.

179. Tyler, *The Chemcraft Story*, 39, 31.

Chapter 4

1. Hickam, *Rocket Boys*, 11–18.

2. Cheng, *Astounding Wonder*; McCurdy, *Space and the American Imagination*, 11–32.

3. For an overview of the widespread popularity of these "Rocket Ranger" television shows and their effects on the market for children's products, see this edited collection: Miller and Van Riper, *1950's "Rocketman" TV Series and Their Fans*, 18.

4. William Gibson, "Olds Rocket 88, 1950," *New Yorker*, June 4, 2012.

5. Hickam, *Rocket Boys*, 281.

6. Three articles in adult general-interest publications that used a comical presentation of spaceman lingo as a colorful hook in the first few paragraphs: Jack Cluett, "Will Your Child Visit the Moon?" *Women's Day*, August 1953, 19; Murray Robinson, "Planet Parenthood," *Collier's Weekly*, January 5, 1952; and "Space Patrol Conquers Kids," *Life*, September 1, 1952.

7. Amos Sewell, "Space Traveller," *Saturday Evening Post*, November 8, 1952; "Junior Space Ranger," *Collier's Weekly*, April 18, 1953.

8. Cluett, "Will Your Child Visit the Moon?," 61.

9. Jack Cluett, "Rocket Bye Baby," *Women's Day*, August 1953, 116.

10. "Four Flight-Tested Space Helmets You Can Make," *Women's Day*, August 1953.

11. "Starmaster Scientific Toys" (Harmonic Reed Corporation, Rosemont, PA, n.d.), SM.

12. Chapter 7, "'Fandom Is Just a Goddamn Hobby': The Industry of Fans and Professionals," in Cheng, *Astounding Wonder*; Svilpis, "Authority, Autonomy, and Adventure." "During lunch hour," Asimov wrote of the year he discovered pulp science fiction, "we would sit on the curb in front of the school, and to anywhere from two to ten eager listeners I would repeat the stories I had read, together with such personal embellishments as I could manage. It increased my pleasure in science fiction, and I discovered that I loved to have an audience." Quotation from his autobiography, *In Memory Yet Green* (New York: Doubleday, 1979), 106, quoted in Cheng, *Astounding Wonder*, 241.

13. Berger, "Love, Death, and the Atomic Bomb," 280.

14. "Pulp Magazines Called a Menace; Hartford Teacher Declares They Threaten Morals of High School Students. She Suggests Antidotes, Tells National Council of Teachers at Boston Mental Effort Must Be Stimulated," *New York Times*, November 29, 1936.

15. Sullivan, "American Young Adult Science Fiction," 21. Sullivan points to the year of *Rocket Ship Galileo*'s publication as a definitive turning point in the legitimization of science fiction for young adult readers.

16. "The Forthcoming Juveniles Will Reflect Changes in Children's Interests," *Publishers' Weekly*, July 29, 1950. 413.

17. *Books for Young People* (1950); *Books for Young People* (1953); Rouyer, "How Did YA Become YA?" A New York Public Library librarian, Margaret Scoggin, coined the term "young adult" when she renamed her long-running column for *Library Journal* in 1944,

changing it from "Books for Older Boys and Girls" to "Books for Young Adults." The genre did not emerge into its full strength until the 1970s.

18. At least one reviewer juxtaposed the Winston books with Heinlein's juvenile oeuvre, with the former emerging the worse for the comparison: "Characterization, motivation, and even description are meager, if at all present." Villiers Gerson, "Out in Space," *New York Times*, June 28, 1953.

19. Svilpis, "Authority, Autonomy, and Adventure." Svilpis's article contains a more complete overview of the science fiction and science-fiction-adjacent work produced for adolescents during this period. To these examples, I would add Raymond Abrashkin and Jay Williams' *Danny Dunn* series; Nelson Bond, *The Remarkable Exploits of Lancelot Biggs, Spaceman* (1950); Bertrand Brinley, *The Mad Scientists' Club* (1965); Eleanor Cameron and Robert Henneberger's *The Wonderful Flight to the Mushroom Planet* (1954) and sequel (1956); E. C. Elliott's (Reginald Alec Martin) *Kemlo* series (1954–63); Ellen MacGregor, *Miss Pickerell Goes to Mars* (1951) and sequels (1953, 1953, 1954); William Morrison, *Mel Oliver and Space Rover on Mars* (1954); Chad Oliver, *Mists of Dawn* (1952); the books of Andre (Alice) Norton, including *Star Man's Son, 2250 A.D.* (1952); and Carey Rockwell, *Space Pioneers* (1953).

20. Franklin, *Robert A. Heinlein*, 14. Heinlein acted as a consultant on "Destination Moon," but its plot was substantially altered from *Rocket Ship Galileo's* initial outline, and its protagonists were grown men; the film was intended for an adult audience. Heinlein was not officially associated with "Tom Corbett: Space Cadet" and wrote to his editor and her assistant that although the show was popular, he did not think Scribner's should try to promote *Space Cadet* by associating it with the serial, which he found "moronic." Robert A. Heinlein to Alice Dalgliesh and Virginia Fowler, January 5, 1951, Box 333, HA.

21. Bechtel, "Imagination's Other Place," 180. For biographical information on Bechtel, see "Guide to the Louise Seaman Bechtel Papers, 1877–1980 (bulk 1913–1980)," Vassar College Libraries, http://specialcollections.vassar.edu/collections/findingaids/b /bechtel_louise_seaman.html (accessed June 4, 2015).

22. Helene Frye, "What about Science Books?" *Top of the News*, November 1964, 30.

23. Ryan, ed., *Across the Space Frontier*.

24. Agnes Krarup, "Books and Defense," *Library Journal*, February 15, 1959, 1–2. William Gibson remembered a similar scene. As a child, he argued with an "Air Force Man, a visitor to our home, who made mock of my Willy Ley book [*The Conquest of Space*]. I knew he was wrong when he said that space travel would never happen. And I was right, at least in the relatively short term, just a few years off from Sputnik. I was a native, I felt unquestioningly, of Tomorrow." Gibson, "Olds Rocket 88, 1950."

25. Joseph Gallant, ""Literature, Science, and the Manpower Crisis." *Science*, April 26, 1957.

26. Geoff Conklin, "What Is Good Science Fiction?" *Junior Libraries*, April 15, 1958, 18.

27. Learned T. Bulman, "Using Science Fiction as Bait," *Library Journal*, December 15, 1955, 8.

28. Isaac Asimov, "The By-Product of Science Fiction," *AIBS Bulletin* 7, no. 1 (January 1, 1957): 25–27.

29. Alexei Panshin, "Heinlein's Child," http://www.enter.net/~torve/critics/child .html (accessed May 24, 2009). I should note that Panshin later wrote a work of literary criticism about Heinlein's oeuvre, *Heinlein in Dimension* (1968), that created such tension between him and Heinlein that Heinlein sued to stop its publication; the controversy over this book, which eventually won a Hugo Award, is still ongoing in fan communities (see, for example, "What Is It with Heinlein Fans and Panshin?" Straight Dope Message Board, n.d., http://boards.straightdope.com/sdmb/showthread.php?t=168279 (accessed April 12, 2011).

30. Sullivan, "Heinlein's Juveniles," 64.

31. The label "juvenile" was used for books intended for young adults before the 1970s, when "young adult" or "YA" became common. See Mendlesohn, *Inter-galactic Playground*, 5.

32. On Heinlein's public image as libertarian, see Stan Lehr and Louis Rossetto, "The New Right Credo—Libertarianism," *New York Times*, January 10, 1971. This article, written by two seniors at Columbia, links Heinlein to Ayn Rand, Goldwater speechwriter Karl Hess, and Murray Rothbard, editor of the newsletter *Libertarian Forum*; the piece points to the 1966 Heinlein novel about a community on the moon that revolts against a dictatorship, *The Moon Is a Harsh Mistress*, as a key text.

33. Sullivan, "Heinlein's Juveniles: Growing Up in Outer Space," 65.

34. Breines, *Young, White, and Miserable*; Ladd-Taylor and Umansky, *"Bad" Mothers*; Wylie, *Generation of Vipers*.

35. Patterson, *Robert A. Heinlein*, 27.

36. Ibid., 315; Terry, "'Momism' and the Making of Treasonous Homosexuals," 175.

37. Patterson, *Robert A. Heinlein*, 28.

38. Robert Heinlein, "Sworn Statement of Robert A. Heinlein concerning Robert Cornog," May 17, 1945, Box 306, HA.

39. Franklin, *Robert A. Heinlein*, 4–5.

40. A complete list of Heinlein's Scribner's titles, with publication dates: *Rocket Ship Galileo*, 1947; *Space Cadet*, 1948; *Red Planet*, 1949; *Between Planets*, 1951; *The Rolling Stones*, 1952; *Farmer in the Sky*, 1953; *Starman Jones*, 1953; *The Star Beast*, 1954; *Tunnel in the Sky*, 1955; *Time for the Stars*, 1956; *Citizen of the Galaxy*, 1957; *Have Space Suit—Will Travel*, 1958.

41. Robert Heinlein to Lurton Blassingame, January 29, 1945, Box 331, HA; Robert Heinlein to George, February 12, 1959, Box 333, HA.

42. On the place of the Stratemeyer Syndicate in producing books for earlier generations of teenage boys, see Baxter, *The Modern Age*. Robert A. Heinlein to Lurton Blassingame, March 4, 1949, Box 331, HA.

43. Robert Heinlein, "Tomorrow the Moon," 1947, Box 333, HA.

44. Ellen Lewis Buell, "There's Fact and Fancy—And the Horse Is Still King," *New York Times*, November 12, 1950.

45. Richard S. Alm, "The Development of Literature for Adolescents," *School Review* 64, no. 4 (April 1956): 177.

46. Association for Library Service to Children, "Newbery Medal and Honor Books, 1922–Present."

47. "You Meet Such Interesting People," *Publishers' Weekly*, March 21, 1960, 37.

48. Dalgliesh, "Improvement in Juvenile Books during the Last Ten Years," *Publishers' Weekly*, October 25, 1930, 1471.

49. Eddy, *Bookwomen*, 156–57.

50. In some cases, Dalgliesh sent Heinlein's manuscripts to such figures as Helen Ferris, the editor of the Junior Literary Guild, or Margaret Scoggin, who hosted the WMCA radio show *Young Book Reviewers*, and then wrote back to Heinlein quoting liberally from the comments of these respected outsiders as support for her own opinions. Alice Dalgliesh to Robert Heinlein, December 26, 1946, Box 333, HA; Alice Dalgliesh to Lurton Blassingame, March 18, 1949, Box 333, HA.

51. Heinlein, *Rocket Ship Galileo*, 15–16.

52. Other objects of children's culture produced during this time also featured the informal science education that took place when a child encountered a "real" (usually male) scientist. The television show "Mr. Wizard" (1951–65), for example, revolved around the encounters between the avuncular Wizard and the children living in his neighborhood, while the Danny Dunn series, by Raymond Abrashkin and Jay Williams, featured a boy who forms a bond with a scatter-brained professor, Euclid Bullfinch, and learns his science through helping Bullfinch with his experiments. These learning situations abstracted science teaching from the school setting; as in Heinlein's books, the message is that one way to truly "get" science was through one-on-one contact with a male teacher with practical experience.

53. Robert Heinlein to Alice Dalgliesh, February 17, 1959, Box 333, HA (emphasis in the original).

54. Ibid.

55. Robert Heinlein to Alice Dalgliesh, April 30, 1957, Box 333, HA.

56. Heinlein, *Have Space Suit—Will Travel*, 8–13.

57. Robert Heinlein to Alice Dalgliesh, May 30, 1947, Box 333, HA.

58. Heinlein, *Space Cadet*, 75.

59. Ibid., 76.

60. Heinlein loved Horatio Alger stories as a child and wrote that his juveniles were plotted in the same fashion as Alger's rags-to-riches stories, except that the trajectory of his protagonists is not from poor to rich but from boy to man. Robert Heinlein to Alice Dalgliesh, February 3, 1959, Box 333, HA.

61. Heinlein, "Ray Guns and Rocket Ships," 1190.

62. Robert A. Heinlein to Lurton Blassingame, March 4, 1949, Box 331, HA (emphasis in original).

63. Ibid.

64. Robert A. Heinlein to Lurton Blassingame, March 15, 1949, Box 331, HA.

65. Heinlein to Blassingame, March 4, 1949.

66. Robert Heinlein to Lurton Blassingame, March 16, 1946, Box 331, HA.

67. John to Robert Heinlein, February 10, 1949, Box 306, HA.

68. Heinlein to George, February 12, 1959.

69. Robert Heinlein to Anna Lyle, April 10, 1951, Box 306, HA.

70. Ibid., 76.

71. Heinlein to Blassingame, March 4, 1949.

72. Ibid.

73. Robert Heinlein to Alice Dalgliesh, May 1, 1957, Box 333, HA.

74. Dalgliesh, "To Light a Candle," *Publishers' Weekly*, April 28, 1943, 1735.

75. Eakin, "Trends in Children's Literature," *Library Quarterly* 25, no. 1 (January 1, 1955): 51.

76. Heinlein to Blassingame, March 16, 1946.

77. Rahn, "'Like a Star through Flying Snow,'" 316.

78. Robert Heinlein to Alice Dalgliesh, May 2, 1955, Box 333, HA.

79. Heinlein, *Red Planet*, 170.

80. Robert Heinlein to Lurton Blassingame, March 24, 1949, Box 331, HA.

81. Robert Heinlein to Alice Dalgliesh, May 13, 1954, Box 333, HA.

82. Ibid., April 11, 1958.

83. Ibid., December 24, 1958.

84. Spigel, "From Domestic Space to Outer Space: The 1960s Fantastic Family Sitcom," in *Welcome to the Dreamhouse*, 141–83.

85. Learned T. Bulman to Alice Dalgliesh, August 30, 1954, Box 333, HA.

86. Robert Heinlein to Learned T. Bulman, September 7, 1954, Box 333, HA.

87. Robert Heinlein to Alice Dalgliesh, September 7, 1954, Box 333, HA; ibid., April 2, 1956.

88. Heinlein to Dalgliesh, May 13, 1954.

89. Bulman, "Using Science Fiction as Bait," 8. In 1958, after *Have Space Suit—Will Travel* was published, Bulman wrote Heinlein a congratulatory letter, calling the book "excellent" and adding that he agreed with the book's assessment of contemporary schooling. Bulman to Heinlein, October 24, 1958, Box 333, HA.

90. Frank, "Women in Heinlein's Juveniles"; Russ, "Images of Women in Science Fiction."

91. Robert Heinlein to Virginia Fowler, August 1, 1949, Box 333, HA.

92. Robert Heinlein to Alice Dalgliesh, April 11, 1951, Box 333, HA.

93. Patterson, *Robert A. Heinlein*, 308.

94. Robert Heinlein to Cal Laning, December 2, 1948, Box 306, HA.

95. Heinlein, *Podkayne of Mars*, 126–27.

96. Franklin, *Robert A. Heinlein*, 144.

97. This book drew extensive criticism from more left-wing sectors of the science fiction community. In 1978, for example, science fiction writer Michael Moorcock wrote, in an article titled "Starship Stormtroopers," that the book was "pure debased Ford out

of Kipling, setting the pattern for Heinlein's more ambitious paternalistic, xenophobic (but equally sentimental) stories," such as *Farnham's Freehold* (1964), a futuristic fantasy in which cruel black people rule the world and a white hero fights them. Michael Moorcock, "Starship Stormtroopers," December 24, 2002, http://flag.blackened.net/liberty /moorcock.html (accessed April 6, 2011).

98. Gilbert, *A Cycle of Outrage*.

99. Alice Dalgliesh to Robert Heinlein, February 11, 1959, Box 333, HA (emphasis in the original).

100. Heinlein to Dalgliesh, February 17, 1959.

101. Ibid.

102. Stearns, *Anxious Parents*, 218.

103. Robert Heinlein to Ruth Clement Boyer, May 22, 1947, Box 333, HA.

104. John Reubens, "The Gun Is Pointed at Our Heads," n.d., Box 209, HA.

105. William J. Barker, "Hello, Dog?" n.d., Box 209, HA.

106. A. E. Housman, "IX," from *Last Poems*, 1922; Horace Gregory, "Voices of Heroes," published in *Free World*, April 1942; Horace, "The Odes," book III:II. Wilfred Owen, British poet of World War I, appropriated the Horace line to title one of his most famous poems, which condemned the idea of glorious patriotic death altogether; Heinlein did not note the irony. Suggested titles in Robert Heinlein to William McMorris, August 20, 1959, Box 332, HA.

107. Heinlein to Dalgliesh, February 3, 1959.

108. Franklin, *Robert A. Heinlein*, 20–21.

Chapter 5

1. Jack Rosenbaum, "Subtle Psychology," *San Francisco Progress*, July 1, 1977.

2. Carol Pogash, "An Official Good Time at Exploratorium," *San Francisco Examiner*, n.d.

3. "Doing What Classrooms Can't," *Mosaic*, Spring 1973, 1.

4. Joan Baldwin, "Exploratorium," *Motorland/CSAA*, December 1973, 22.

5. Carol A. Crotta, "Exploratorium," *Los Angeles Herald Examiner*, January 14, 1979.

6. "Shrieking with Delight at 'a Science Fair Gone Blissfully Mad,'" *Peninsula Times Tribune*, April 24, 1981; Thomas J. O'Neil, "Science in San Francisco's Palace of Perception," *Museum Magazine*, August 1980; Ed Levitt, "A Scientific Disneyland in S.F.," *Tribune*, May 31, 1979.

7. Eleanor Kaplan, "The Exploratorium: New San Francisco Landmark," *Junior League of Oakland East Bay*, December 1976.

8. Frank Oppenheimer, "Everyone Is You—or Me," 1976, 2, Carton 5, Folder 24, ER.

9. Mona Gable, "Coming to Grips with Science," *Focus*, March 1982, 18.

10. Robert Larkins and Stephanie Peirolo, "Perspectives on Science in a Nuclear Age," *Up Front*, Winter 1984, 4.

11. On the language of self-actualization, see Jessica Grogan, *Encountering America:*

Humanistic Psychology, Sixties Culture, and the Shaping of the Modern Self (New York: Harper Perennial, 2012).

12. Miller, "Public Understanding of, and Attitudes Toward, Scientific Research." Miller cites results of the National Science Foundation's Science and Engineering Indicators studies, which expanded on an earlier set of surveys carried out by the National Association of Science Writers. When asked if they agreed with the statement, "Science and technology are making our lives healthier, easier, and more comfortable," 94 percent of respondents agreed in 1957; 81 percent in 1979; 90 percent in 1999. Asked if they agreed with the statement, "Science makes our way of life change too fast," 40 percent agreed in 1957; in 1979, 50 percent; in the 1990s, back to 40 percent. Asked, "Have the benefits of scientific research outweighed the harmful results?" in 1981, 70 percent agreed; in 1999, 75 percent agreed.

13. Spigel, "Outer Space and Inner Cities: African-American Responses to NASA," in *Welcome to the Dreamhouse*, 141–83; Tribbe, *No Requiem for the Space Age*.

14. On seventies environmental rhetoric and negativity toward science, see Buell, *From Apocalypse to Way of Life*; Dunaway, "Gas Masks, Pogo, and the Ecological Indian"; Egan, *Barry Commoner and the Science of Survival*; Gottlieb, *Forcing the Spring*; Reich, "From the Spirit of St. Louis to the SST"; Rome, *The Genius of Earth Day*; and Sabin, *The Bet*.

15. Viet Nam Generation, Inc., "Port Huron Statement," *The Sixties Project*, 1962, http://www2.iath.virginia.edu/sixties/HTML_docs/Resources/Primary/Manifestos /SDS_Port_Huron.html (accessed October 5, 2012).

16. Goodman, *Growing Up Absurd*, 105–6.

17. D. Philip Baker, Howard E. Smith, and Thomas G. Aylesworth, "The Juvenile Book Editor: An Interview," *Library Trends*, April 1974, 433–41.

18. Danny Miller, "I Heart Amy Carter," *Huffington Post*, May 25, 2011, http://www .huffingtonpost.com/danny-miller/i-heart-amy-carter_b_14402.html (accessed June 18, 2015).

19. I have written about the discourse surrounding "The Day After"; see Onion, "Honey, You're Scaring the Kids."

20. Quoted in Rome, *The Genius of Earth Day*, 38.

21. Tom Kubald to Gaylord Nelson, February 3, 1970, GN.

22. DeMuth's words quoted in a press release from the National Education Association, "National Environmental Teach-In Captures Imaginations of Students, Teachers," April 7, 1970, GN.

23. Rome, *The Genius of Earth Day*, 106.

24. Ammon, "Unearthing 'Benny the Bulldozer,'" 311. Ammon points to Edith Thacher Hurd and Clement Hurd's *Benny the Bulldozer* as an example of what she calls the "culture of clearance" in postwar children's books. This is a particularly telling example with which to see change in attitudes over time, because Edith Thacher Hurd later published a book that was an explicit critique of overdevelopment (Hurd, *Wilson's World* [1971]). For the friendly snowplow, see Virginia Lee Burton, *Katy and the Big Snow*.

25. Other sad early-1970s picture books include Foreman, *Dinosaurs and All That Rubbish*; Hurd, *Wilson's World*; Martin and Ells, *Spoiled Tomatoes*; Parnall, *The Mountain*; and Peet, *The Wump World*.

26. George Zebrowski, "The Firebird," *Ranger Rick*, October 1979.

27. Plevin, "Still Putting Out 'Fires,'" 168–82.

28. Ruth Reichl, "Dreams of a Mad Scientist," *New West*, July 17, 1978.

29. On J. Robert Oppenheimer's public significance, see the work of David Hecht, *Storytelling and Science*, "The Atomic Hero," and "A Nuclear Narrative." The Frank Oppenheimer biography is by Cole, *Something Incredibly Wonderful Happens*.

30. On this distinction, see Steven Shapin, "Uncle of the Bomb," *London Review of Books*, September 23, 2010.

31. Ibid.

32. Sam Maddox, "Life in the Shadows," *Sunday Camera Magazine*, March 24, 1985; Shapin, "Uncle of the Bomb."

33. Nathanson, "The Oppenheimer Affair."

34. Paul Preuss, "On the Blacklist," *Science*, June 1983, 38. On the NSF's scientist-led curriculum projects, see Rudolph, *Scientists in the Classroom*.

35. Gable, "Coming to Grips with Science."

36. Quoted in Shapin, "Uncle of the Bomb," 13.

37. The Science Museum in South Kensington, London; the Deutsches Museum in Munich; and the Palais de la Découverte in Paris.

38. Frank Oppenheimer, "Frank Oppenheimer—Biography," n.d., Carton 7, Folder 22, ER; "Biography: Dr. Frank Oppenheimer" (Association of Science-Technology Centers, May 1981), Carton 7, Folder 23, ER; "Frank Oppenheimer—Chronology," n.d., Carton 7, Folder 22, ER; Frank Oppenheimer, "The Exploratorium: A Playful Museum Combines Perception and Art in Science Education." *American Journal of Physics* 40 (July 1972): 978–84; Gwendolyn Evans, "An Explorer's Reward," *San Francisco Chronicle*, August 28, 1974.

39. Rader and Cain, *Life on Display*, 213.

40. Reichl, "Dreams of a Mad Scientist," 85.

41. "The Exploratorium as a Model for Museums," n.d., Carton 21, Folder 6, ER; Rader and Cain, *Life on Display*, chap. 6, "The Exploratorium Effect: Redefining Relevance and Interactive Display, 1969–1980."

42. Judith Dunham, "Conjunctions of Art and Science," *Art Week*, May 6, 1978, 1.

43. O'Neil, "Science in San Francisco's Palace of Perception," 37.

44. Frank Oppenheimer Interview for *NOVA*/Exploratorium Documentary, interview by John Else, July 2, 1981, Carton 7, Folder 31, ER.

45. Steven Shapin identifies Oppenheimer's philosophy as a throwback to the scientism of figures such as John Dewey. Shapin, "Uncle of the Bomb."

46. Frank Oppenheimer, "Frank Oppenheimer Interview," October 10, 1981, 4, Carton 21, Folder 51, ER.

47. Preuss, "On the Blacklist," 35.

48. "'The Sentimental Fruits of Science,' Dr. Frank Oppenheimer's Acceptance Speech for the Oersted Medal Awarded by the American Association of Physics Teachers, San Antonio, Texas, 31 January 1984," *American Journal of Physics* 52, no. 8 (August 1984): 686.

49. Frank Oppenheimer Interview for *NOVA*/Exploratorium Documentary.

50. Frank Oppenheimer, "Exploring," 1983, 4, Carton 5, Folder 45, ER.

51. Frank Oppenheimer, "Schools Are Not for Sightseeing," 1970, Carton 5, Folder 17, ER.

52. Frank Oppenheimer, "Exhibit Conception and Design" (paper presented at a meeting of International Commission on Science Museums, Monterey, Mexico, 1980), 4.

53. Linda Dackman, "Invisible Aesthetic: A Somewhat Humorous, Slightly Profound Interview with Frank Oppenheimer," 1983, Carton 7, Folder 37, ER.

54. Sherwood Davidson Kohn, "It's OK to Touch at the New Hands-On Exhibits," *Smithsonian*, September 1978, 81.

55. Shirley Boes Neill, "Exploring the Exploratorium," *American Education*, December 1978, 11.

56. Ogata, *Designing the Creative Child*.

57. See more on the idea of "islanding" in Chapter 1. Gillis, "The Islanding of Children."

58. On children's liberation and children's rights, see Archard, *Children*, 64–68. Texts of the 1970s children's liberation movement include Farson, *Birthrights*; Gross and Gross, *The Children's Rights Movement*; and Holt, *Escape from Childhood*.

59. Edwin Kiester, "Please Touch," *Parents*, September 1978.

60. Levitt, "A Scientific Disneyland in S.F."

61. Frank Oppenheimer, "Adult Play," *Exploratorium Magazine*, 1980; Frank Oppenheimer, "Curiosity," *San Francisco Sunday Examiner & Chronicle*, September 25, 1983.

62. Oppenheimer, "Adult Play," 2.

63. Cole, *Something Incredibly Wonderful Happens*, 10.

64. Even in 1981, people under eighteen could enter the museum free.

65. "Biography: Dr. Frank Oppenheimer" (Association of Science-Technology Centers, May 1981), Carton 7, Folder 23, ER.

66. K. C. Cole, "Exploratorium," n.d., Carton 21, Folder 32, ER.

67. Frank Oppenheimer, "Exploring," 8.

68. Pat Angle, "The Exploratorium," *IJ Living*, January 24, 1981, 41.

69. "Doing What Classrooms Can't," 2.

70. "Biography: Dr. Frank Oppenheimer." On the place of philosophical toys in nineteenth-century popular science, see Morus, "Seeing and Believing Science," and Wade, "Philosophical Instruments and Toys."

71. Neill, "Exploring the Exploratorium," 10.

72. Cole, "Exploratorium."

73. Syder, "'Shaken Out of the Ruts of Ordinary Perception.'"

74. "'The Sentimental Fruits of Science,'" 686.

75. Oppenheimer, "The Exploratorium," 982.

76. Oppenheimer, "Exploring," 7.

77. James O. Clifford, "The Museum's a Touching Experience," *South Bay Daily Breeze*, May 12, 1975.

78. Frank Oppenheimer Interview for *NOVA*/Exploratorium Documentary.

79. Ibid.

80. Frank Oppenheimer, "Museums: A Versatile Resource for Learning and Pleasure," n.d., 2, Carton 5, Folder 57, ER.

81. "Young Explainers Learn and Earn," *Cultural Post*, September 1975.

82. Dan Borsuk, "Minority Students Learn through Teaching," *San Francisco Progress*, November 25, 1980.

83. Oppenheimer, "The Exploratorium," 983.

84. Ibid., 980.

85. Frank Oppenheimer Interview for *NOVA*/Exploratorium Documentary.

86. Dackman, "Invisible Aesthetic."

87. "Doing What Classrooms Can't," 2.

88. Evelyn Shaw, "The Exploratorium," *Curator* 15, no. 1 (1972): 39.

89. Neill, "Exploring the Exploratorium," 8.

90. Cole, "Exploratorium."

91. Jenkins, "Dennis the Menace."

92. Ketcham, *Dennis the Menace*.

93. Perlman, a figure of Oppenheimerian vigor who was still reporting stories for the *San Francisco Chronicle* in 2013 at the age of ninety-four, covered the Exploratorium's move to a more modern building on the Embarcadero that year. Maria L. La Ganga, "Science Writer Is Quite the Specimen Himself: He's 94," *Los Angeles Times*, February 21, 2013.

94. The current adult admission fee is $29; for children under eighteen, $24. "Buy Tickets," *Exploratorium*, http://www.exploratorium.edu/visit/tickets (accessed June 25, 2015).

Conclusion

1. Teitelbaum, *Falling Behind?*, 3. An example of a boom and bust would be the increase in funding for physics after Sputnik, which sputtered in the late 1960s and early 1970s, leaving many newly minted physicists out of work (a phenomenon David Kaiser writes about in his book *How the Hippies Saved Physics*).

2. Charette, "The STEM Crisis Is a Myth."

3. Louv, *Last Child in the Woods*; Hanna Rosin, "The Overprotected Kid," *Atlantic*, March 19, 2014, http://www.theatlantic.com/features/archive/2014/03/hey-parents -leave-those-kids-alone/358631/ (accessed March 21, 2014); Lenore Skenazy, "Teach Kids Safety Rules, but Don't Keep Them Inside," *Daily Beast*, October 24, 2012, http:// www.thedailybeast.com/articles/2012/10/24/lenore-skenazy-teach-kids-safety-rules -but-don-t-keep-them-inside.html (accessed October 24, 2012).

4. Malcolm W. Browne, "A Goodbye to Adventures Gone By," *New York Times*, January 29, 1980.

5. Silberman, Steve, "Don't Try This at Home," *Wired*, June 2006.

6. Eileen Pollack, "Why Are There Still So Few Women in Science?" *New York Times*, October 3, 2013. Pollack's book on the topic is *The Only Woman in the Room*.

7. Bell, "Science as 'Horrible.'"

8. Wannamaker, *Boys in Children's Literature and Popular Culture*.

9. Kate Clancy, "Girls with Toys: This Is What Real Scientists Look Like," *Slate*, May 18, 2015, http://www.slate.com/articles/health_and_science/science/2015/05/girls _with_toys_on_twitter_feminist_hashtag_shares_images_of_women_doing.html (accessed June 5, 2015).

10. Marie-Claire Shanahan, "Inspiration from Bassist Victor Wooten Shows Me a New Way to Deal with My 'Child-as-Scientist' Frustrations," *Boundary Vision*, August 13, 2012, http://boundaryvision.com/2012/08/13/inspiration-from-bassist-victor-wooten-shows -me-a-new-way-to-deal-with-my-child-as-scientist-frustrations/ (accessed August 13, 2012).

11. Ariel Waldman, "Adults Are the Future," *ArielWaldman.com*, February 13, 2014, http://arielwaldman.com/2014/02/13/adults-are-the-future/ (accessed May 14, 2015).

Bibliography

Archival Sources

Berkeley, California
  University of California, Bancroft Library, Exploratorium Records, BANC MSS
    87/148 c
Brooklyn, New York
  Brooklyn Children's Museum, Brooklyn Children's Museum Archives
Madison, Wisconsin
  Gaylord Nelson Archives, Wisconsin Historical Society, http://nelsonearthday.net
    /about/whs.htm
Philadelphia, Pennsylvania
  Chemical Heritage Foundation Archives
Princeton, New Jersey
  G. Edward Pendray Papers, Princeton University Archives
Rochester, New York
  Archives of the Strong National Museum of Play
Santa Cruz, California
  University of California, Heinlein Prize Trust, Robert A. Heinlein Archives
Washington, D.C.
  Smithsonian Institution Archives, Record Unit 7091

Newspapers and Periodicals

Advertising & Selling
AIBS Bulletin
American Education
American Museum Journal
Art Week
Books for Young People
  (*Books for the Teen Age*)
Boys' Life
Business Week
Chemcraft Science Magazine

Chemistry   *New Yorker*
Childhood   *New York Times*
Child Labor Bulletin   *Parents*
Children's Museum News   *Peninsula Times Tribune*
Collier's Weekly   *Playthings*
Commentary   *Popular Science Monthly*
Commonweal   *Ranger Rick*
Cultural Post   *Reader's Digest*
Education Week (online)   *Saga Magazine*
80 Beats (*Discover Magazine* blog)   *San Francisco Chronicle*
Engineers' Bulletin   *San Francisco Examiner*
Focus   *San Francisco Progress*
4-H Horizons   *San Francisco Sunday Examiner &*
Free World   *Chronicle*
Hygeia   *Saturday Evening Post*
IJ Living   *School Life*
Independent   *School Review*
Junior Eagle   *School Science and Mathematics*
Junior League of Oakland East Bay   *Science News-Letter*
Junior Libraries   *Science Page*
Life   *Science Teacher*
Literary Digest   *Scientific American*
London Review of Books   *Slate Magazine*
Los Angeles Herald Examiner   *Smithsonian*
Los Angeles Times   *South Bay Daily Breeze*
Merchandising News   *Sunday Camera Magazine*
Mosaic   *Top of the News*
Motorland/CSAA   *Up Front*
Museum Magazine   *Wired*
New West   *Women's Day*

Books, Journal Articles, and Unpublished Papers

Adas, Michael. *Dominance by Design: Technological Imperatives and America's Civilizing Mission.* Cambridge, Mass.: Belknap Press of Harvard University Press, 2006.

Adorno, Theodor. *The Authoritarian Personality.* New York: Harper, 1950.

Alcorn, Aaron L. "Flying into Modernity: Model Airplanes, Consumer Culture, and the Making of Modern Boyhood in the Early Twentieth Century." *History & Technology* 25, no. 2 (June 2009): 115–46.

Alexander, Edward P. *The Museum in America: Innovators and Pioneers.* Walnut Creek, Calif.: AltaMira Press, 1997.

Al-Gailani, Salim. "Magic, Science and Masculinity: Marketing Toy Chemistry Sets." *Studies in History and Philosophy of Science Part A* 40, no. 4 (December 2009): 372–81.

Ammon, Francesca Russello. "Unearthing 'Benny the Bulldozer': The Culture of Clearance in Postwar Children's Books." *Technology and Culture* 53, no. 2 (2012): 306–36.

Angier, Natalie. *The Canon: A Whirligig Tour of the Beautiful Basics of Science.* Boston: Houghton Mifflin, 2007.

Antler, Joyce. *Lucy Sprague Mitchell: The Making of a Modern Woman.* New Haven: Yale University Press, 1987.

Archard, David. *Children: Rights and Childhood.* New York: Routledge, 1993.

Armitage, Kevin C. "Bird Day for Kids: Progressive Conservation in Theory and Practice." *Environmental History* 12, no. 3 (July 1, 2007): 528–51.

———. *The Nature Study Movement: The Forgotten Popularizer of America's Conservation Ethic.* Lawrence: University Press of Kansas, 2009.

Association for Library Service to Children, "Newbery Medal and Honor Books, 1922–Present." Accessed April 17, 2011. http://www.ala.org/ala/mgrps/divs/alsc/awards grants/bookmedia/newberymedal/newberyhonors/newberymedal.cfm#30s.

Atkinson, Eleanor. *The How and Why Library: Little Questions That Lead to Great Discoveries.* Vol. 3. Cleveland, Ohio: L. J. Bullard, 1934.

Ball, Terence. "The Politics of Social Science in Postwar America." In *Recasting America: Culture and Politics in the Age of Cold War,* edited by Lary May, 76–92. Chicago: University of Chicago Press, 1989.

Barrow, Mark V. "The Specimen Dealer: Entrepreneurial Natural History in America's Gilded Age." *Journal of the History of Biology* 33, no. 3 (Winter 2000): 493–534.

Baxter, Kent. *The Modern Age: Turn-of-the-Century American Culture and the Invention of Adolescence.* Tuscaloosa: University of Alabama Press, 2008.

Bechtel, Louise Seaman. "Imagination's Other Place." In *Books in Search of Children; Speeches and Essays,* edited by Virginia Haviland, 180–84. New York: Macmillan, 1969.

Bederman, Gail. *Manliness and Civilization: A Cultural History of Gender and Race in the United States, 1880–1917.* Chicago: University of Chicago Press, 1995.

Bell, Alice R. "Science as 'Horrible': Irreverent Deference in Science Communication." *Science as Culture* 20, no. 4 (2011): 491–512.

Benedict, Barbara M. *Curiosity: A Cultural History of Early Modern Inquiry.* Chicago: University of Chicago Press, 2001.

Berger, A. I. "Love, Death, and the Atomic Bomb: Sexuality and Community in Science Fiction, 1935–55 (L'Amour, la Mort et la Bombe Atomique: Sexualité et Communauté dans la SF de 1935 à 1955)." *Science Fiction Studies* 8, no. 3 (1981): 280–96.

Berger, Joseph. *The Young Scientists: America's Future and the Winning of the Westinghouse.* Reading, Mass.: Addison-Wesley, 1994.

Besser, Marianne. *Growing Up with Science.* New York: McGraw-Hill, 1960.

Bowler, Peter J. *Science for All: The Popularization of Science in Early Twentieth-Century Britain.* Chicago: University of Chicago Press, 2009.

Brandwein, Paul Franz. *The Gifted Student as Future Scientist: The High School Student and His Commitment to Science.* New York: Harcourt, Brace, 1955.

Breines, Wini. *Young, White, and Miserable: Growing Up Female in the Fifties*. Boston: Beacon Press, 1992.

Buell, Frederick. *From Apocalypse to Way of Life: Environmental Crisis in the American Century*. New York: Routledge, 2003.

Burnham, John C. *How Superstition Won and Science Lost: Popularization of Science and Health in the United States*. New Brunswick, N.J.: Rutgers University Press, 1987.

———. "How the Discovery of Accidental Childhood Poisoning Contributed to the Development of Environmentalism in the United States." *Environmental History Review* 19, no. 3 (October 1, 1995): 57–81.

———. "Why Did the Infants and Toddlers Die? Shifts in Americans' Ideas of Responsibility for Accidents: From Blaming Mom to Engineering." *Journal of Social History* 29, no. 4 (July 1, 1996): 817–37.

Burton, Virginia Lee. *Katy and the Big Snow*. Boston: Houghton Mifflin, 1943.

Bush, Vannevar. *Science, the Endless Frontier: A Report to the President on a Program for Postwar Scientific Research*. Washington, D.C.: National Science Foundation, 1960.

Bycroft, Michael. "Psychology, Psychologists, and the Creativity Movement: The Lives of Method inside and outside the Cold War." In *Cold War Social Science: Knowledge Production, Liberal Democracy, and Human Nature*, edited by Mark Solovey and Hamilton Cravens, 197–214. Basingstoke, England: Palgrave Macmillan, 2012.

Castañeda, Claudia. *Figurations: Child, Bodies, Worlds*. Durham, N.C.: Duke University Press, 2002.

Cattell, Raymond B. "The Personality and Motivations of the Researcher from Measurements of Contemporaries and from Biography." In *Scientific Creativity, Its Recognition and Development. Selected Papers from the Proceedings of the First, Second, and Third University of Utah Conferences: The Identification of Creative Scientific Talent*, edited by Calvin W. Taylor and Frank Barron, 119–31. New York: Wiley, 1963.

Cavallo, Dominick. *Muscles and Morals: Organized Playgrounds and Urban Reform, 1880–1920*. Philadelphia: University of Pennsylvania Press, 1981.

Charette, Robert. "The STEM Crisis Is a Myth." *Institute of Electrical and Electronics Engineers Spectrum*, August 30, 2013. Accessed August 31, 2013. http://spectrum.ieee.org/at-work/education/the-stem-crisis-is-a-myth.

Charles, Mario A. "Bedford-Stuyvesant." In *Encyclopedia of New York City*, edited by Kenneth T. Jackson, 94–95. New Haven, Conn.: Yale University Press, 1995.

Cheng, John. *Astounding Wonder: Imagining Science and Science Fiction in Interwar America*. Philadelphia: University of Pennsylvania Press, 2012.

Chinn, Sarah E. *Inventing Modern Adolescence: The Children of Immigrants in Turn-of-the-Century America*. New Brunswick, N.J.: Rutgers University Press, 2009.

Chorness, Maury H. "An Interim Report on Creativity Research." In *Scientific Creativity, Its Recognition and Development. Selected Papers from the Proceedings of the First, Second, and Third University of Utah Conferences: The Identification of Creative Scientific Talent*, edited by Calvin W. Taylor and Frank Barron, 278–98. New York: Wiley, 1963.

Clark, Clifford Edward. *The American Family Home, 1800–1960*. Chapel Hill: University of North Carolina Press, 1986.

Clowse, Barbara Barksdale. *Brainpower for the Cold War: The Sputnik Crisis and National Defense Education Act of 1958*. Westport, Conn: Greenwood Press, 1981.

Cohen-Cole, Jamie. *The Open Mind: Cold War Politics and the Sciences of Human Nature*. Chicago: University of Chicago Press, 2014.

———. "The Reflexivity of Cognitive Science: The Scientist as Model of Human Nature." *History of the Human Sciences* 18 (2005): 107–39.

Cole, Charles Chester. *Encouraging Scientific Talent: A Study of America's Able Students Who Are Lost to College and of Ways of Attracting Them to College and Science Careers*. New York: College Entrance Examination Board, 1956.

Cole, K. C. *Something Incredibly Wonderful Happens: Frank Oppenheimer and the World He Made Up*. New York: Houghton Mifflin Harcourt, 2009.

Colgrove, James. "'Science in a Democracy': The Contested Status of Vaccination in the Progressive Era and the 1920s." *Isis* 96, no. 2 (June 1, 2005): 167–91.

Conn, Steven. *Museums and American Intellectual Life, 1876–1926*. Chicago: University of Chicago Press, 1998.

Cook, Daniel Thomas. *The Commodification of Childhood: The Children's Clothing Industry and the Rise of the Child Consumer*. Durham, N.C.: Duke University Press, 2004.

Cremin, Lawrence A. *The Transformation of the School: Progressivism in American Education, 1876–1957*. New York: Vintage Books, 1964.

Cross, Gary. "The Cute Child and Modern American Parenting." In *American Behavioral History: An Introduction*, edited by Peter N. Stearns, 19–41. New York: New York University Press, 2005.

———. *Kids' Stuff: Toys and the Changing World of American Childhood*. Cambridge, Mass.: Harvard University Press, 1997.

Dewey, John. *How We Think: A Restatement of the Relation of Reflective Thinking to the Educative Process*. Boston: D. C. Heath, 1933.

Dimock, George. "Priceless Children: Child Labor and the Pictorialist Ideal." In *Priceless Children: American Photographs, 1890–1925*, 7–22. Greensboro, N.C.: Weatherspoon Art Museum, 2001.

Dobson, James C. *Dare to Discipline*. Wheaton, Ill.: Tyndale House Publishers, 1970.

———. *The New Dare to Discipline*. Reissue ed. Wheaton, Ill.: Tyndale Momentum, 2014.

Douglas, Susan J. *Inventing American Broadcasting, 1899–1922*. Baltimore: Johns Hopkins University Press, 1987.

Dunaway, Finis. "Gas Masks, Pogo, and the Ecological Indian: Earth Day and the Visual Politics of American Environmentalism." *American Quarterly* 60 (2008): 67–97.

Eddy, Jacalyn. *Bookwomen: Creating an Empire in Children's Book Publishing, 1919–1939*. Madison: University of Wisconsin Press, 2006.

Egan, Michael. *Barry Commoner and the Science of Survival: The Remaking of American Environmentalism*. Cambridge, Mass.: MIT Press, 2007.

Farson, Richard Evans. *Birthrights*. New York: Macmillan, 1974.

Fass, Paula S. *The Damned and the Beautiful: American Youth in the 1920's*. New York: Oxford University Press, 1977.

Feynman, Richard Phillips. *"Surely You're Joking, Mr. Feynman": Adventures of a Curious Character*. New York: W. W. Norton, 1984.

Foreman, Michael. *Dinosaurs and All That Rubbish*. New York: Crowell, 1973.

Foy, Jessica H., and Thomas J. Schlereth, eds. *American Home Life, 1880–1930: A Social History of Spaces and Services*. Knoxville: University of Tennessee Press, 1992.

Frank, Marietta. "Women in Heinlein's Juveniles." In *Young Adult Science Fiction*, edited by C. W. Sullivan III, 119–30. Westport, Conn.: Greenwood Press, 1999.

Franklin, H. Bruce. *Robert A. Heinlein: America as Science Fiction*. New York: Oxford University Press, 1980.

Frayling, Christopher. *Mad, Bad, and Dangerous? The Scientist and the Cinema*. London: Reaktion, 2005.

Fyfe, Aileen, and Bernard V. Lightman. "Science in the Marketplace: An Introduction." In *Science in the Marketplace: Nineteenth-Century Sites and Experiences*, edited by Aileen Fyfe and Bernard V. Lightman, 1–20. Chicago: University of Chicago Press, 2007.

Gallup, Anna Billings. "A Children's Museum and How Any Town Can Get One." Presentation to the National Education Association of the United States, 1926.

Getzels, J. W., and P. W. Jackson. "The Highly Intelligent and the Highly Creative Adolescent: A Summary of Some Research Findings." In *Scientific Creativity, Its Recognition and Development. Selected Papers from the Proceedings of the First, Second, and Third University of Utah Conferences: The Identification of Creative Scientific Talent*, edited by Calvin W Taylor and Frank Barron, 161–72. New York: Wiley, 1963.

Gilbert, A. C. *The Man Who Lives in Paradise: The Autobiography of A. C. Gilbert*. With Marshall McClintock. Forest Park, Ill.: Heimburger House, 1990.

Gilbert, James. *A Cycle of Outrage: America's Reaction to the Juvenile Delinquent in the 1950s*. New York: Oxford University Press, 1986.

Gillis, John R. "The Islanding of Children—Reshaping the Mythical Landscapes of Childhood." In *Designing Modern Childhoods: History, Space, and the Material Culture of Childhood*, edited by Marta Gutman and Ning De Coninck-Smith, 316–30. New Brunswick, N.J.: Rutgers University Press, 2008.

Glover, Katherine, and Evelyn Dewey. *Children of the New Day*. New York: Appleton-Century, 1934.

Goodman, Paul. *Growing Up Absurd: Problems of Youth in the Organized Society*. New York: Vintage Books, 1960.

Gopnik, Alison. *The Philosophical Baby: What Children's Minds Tell Us about Truth, Love, and the Meaning of Life*. New York: Farrar, Straus and Giroux, 2009.

———. *The Scientist in the Crib: Minds, Brains, and How Children Learn*. 1st ed. New York: William Morrow, 1999.

Gottlieb, Robert. *Forcing the Spring: The Transformation of the American Environmental Movement*. Washington, D.C.: Island Press, 1993.

Gould, Stephen Jay. *Ontogeny and Phylogeny*. Cambridge, Mass.: Belknap Press of Harvard University Press, 1977.

Gregory, Horace. "Voices of Heroes." *Free World* (April 1942): 221–22.

Grogan, Jessica. *Encountering America: Humanistic Psychology, Sixties Culture, and the Shaping of the Modern Self*. New York: Harper Perennial, 2012.

Gross, Beatrice, and Ronald Gross. *The Children's Rights Movement: Overcoming the Oppression of Young People*. Garden City, N.Y.: Anchor Books, 1977.

Gutman, Marta, and Ning de Coninck-Smith. "Introduction: Good to Think With — History, Space, and Modern Childhood." In *Designing Modern Childhoods: History, Space, and the Material Culture of Childhood*, 1–19. New Brunswick, N.J.: Rutgers University Press, 2008.

Hammerton, John. *Child of Wonder: An Intimate Biography of Arthur Mee*. London: Hodder and Stoughton, 1946.

Hansen, Bert. *Picturing Medical Progress from Pasteur to Polio: A History of Mass Media Images and Popular Attitudes in America*. New Brunswick, N.J.: Rutgers University Press, 2009.

Haring, Kristen. *Ham Radio's Technical Culture*. Cambridge, Mass.: MIT Press, 2007.

Harris, Neil. *Cultural Excursions: Marketing Appetites and Cultural Tastes in Modern America*. Chicago: University of Chicago Press, 1990.

Hartman, Andrew. *Education and the Cold War: The Battle for the American School*. New York: Palgrave Macmillan, 2008.

Hartman, Gertrude. *The World We Live In and How It Came to Be: A Pictured Outline of Man's Progress from the Earliest Days to the Present*. New York: Macmillan, 1931.

"Hearings Before a Subcommittee of the Committee on Military Affairs, United States Senate, Seventy-Ninth Congress, Second Session, Pursuant to S. Res. 107 (78th Congress) and S. Res. 146 (79th Congress), Authorizing a Study of the Possibilities of Better Mobilizing the National Resources of the United States. Part 6, Testimony of Science Talent Search Finalists." Washington, D.C.: Government Printing Office, March 5, 1946.

Hecht, David K. "The Atomic Hero: Robert Oppenheimer and the Making of Scientific Icons in the Early Cold War." *Technology and Culture* 49, no. 4 (2008): 943–66.

———. "A Nuclear Narrative: Robert Oppenheimer, Autobiography, and Public Authority." *Biography* 33, no. 1 (2010): 167–84.

———. *Storytelling and Science: Rewriting Oppenheimer in the Nuclear Age*. Amherst: University of Massachusetts Press, 2015.

Heffron, John M. "The Knowledge Most Worth Having: Otis W. Caldwell (1869–1947) and the Rise of the General Science Course." *Science & Education* 4, no. 3 (July 1995): 227–52.

Hein, George E. "Progressive Education and Museum Education: Anna Billings Gallup and Louise Connolly." *Journal of Museum Education* 31, no. 3 (2006): 161–74.

Heinlein, Robert. *Farmer in the Sky*. London: Gollancz, 1975.

———. *Have Space Suit — Will Travel*. New York: Scribner, 1958.

———. *Podkayne of Mars*. New York: G. P. Putnam's Sons, 1963.

———. "Ray Guns and Rocket Ships." *Library Journal*, July 1953.

———. *Red Planet*. London: Gollancz, 1949.

———. *Rocket Ship Galileo*. New York: Scribner, 1947.

———. *Space Cadet*. London: V. Gollancz, 1979.

Herman, Ellen. *The Romance of American Psychology: Political Culture in the Age of Experts*. Berkeley: University of California Press, 1995.

Hersey, John. *The Child Buyer: A Novel in the Form of Hearings before the Standing Committee on Education, Welfare, and Public Morality of a Certain State Senate, Investigating the Conspiracy of Mr. Wissey Jones, with Others, to Purchase a Male Child*. New York: Knopf, 1960.

Hickam, Homer H. *Rocket Boys: A Memoir*. New York: Delacorte Press, 1998.

Higonnet, Anne. *Pictures of Innocence: The History and Crisis of Ideal Childhood*. New York: Thames and Hudson, 1998.

Hilmes, Michele. *Radio Voices: American Broadcasting, 1922–1952*. Minneapolis: University of Minnesota Press, 1997.

Hiner, N. Ray. "Seen but Not Heard: Children in American Photographs." In *Small Worlds: Children and Adolescents in America, 1850–1950*, edited by Elliott West and Paula Petrik, 165–202. Lawrence: University Press of Kansas, 1992.

Hines, Maude. "Review: Playing with Children: What the 'Child' Is Doing in American Studies." *American Quarterly* 61, no. 1 (March 1, 2009): 151–61.

Hoffmann, Banesh. *The Tyranny of Testing*. New York: Crowell-Collier Press, 1962.

Hofstadter, Richard. *Anti-Intellectualism in American Life*. New York: Vintage Books, 1963.

Holt, John Caldwell. *Escape from Childhood*. New York: Ballantine Books, 1975.

Housman, A. E. *Last Poems*. New York: Henry Holt, 1922.

Horace. *The Odes*. Translated by A.S. Kline. Accessed February 17, 2016. http:// www.poetryintranslation.com/PITBR/Latin/HoraceOdesBkIII.htm#anchor _Toc40263847.

Hurd, Edith Thacher. *Wilson's World*. New York: HarperCollins, 1971.

Hurd, Edith Thacher, and Clement Hurt, *Benny the Bulldozer*. Eau Claire, Wisc.: E. M. Hale, 1947.

Innes, Arthur. "Men Who Made the Railways." In *The Book of Knowledge*, edited by Arthur Mee and Holland Thompson, 2:609–16. New York: Grolier Society, 1911.

Jackson, Erika K. "Fit Body, Fit Mind: Scandinavian Youth and the Value of Work, Education, and Physical Fitness in Progressive-Era Chicago." In *Children and Youth during the Gilded Age and Progressive Era*, edited by James Marten, 208–29. New York: New York University Press, 2014.

Jacobson, Lisa. *Raising Consumers: Children and the American Mass Market in the Early Twentieth Century*. New York: Columbia University Press, 2004.

James, Allison, and Alan Prout. "Re-presenting Childhood: Time and Transition in the Study of Childhood." In *Constructing and Reconstructing Childhood: Contemporary*

Issues in the Sociological Study of Childhood, edited by Allison James and Alan Prout, 230–50. London: RoutledgeFalmer, 1997.

Jenkins, Henry. "Dennis the Menace: 'All-American Handful.'" In *The Revolution Wasn't Televised: Sixties Television and Social Conflict*, edited by Lynn Spigel and Michael Curtin, 119–38. New York: Routledge, 1997.

Jenks, Chris. *Childhood*. London: Routledge, 1996.

Kaiser, David. *How the Hippies Saved Physics: Science, Counterculture, and the Quantum Revival*. New York: W. W. Norton, 2012.

———. "The Postwar Suburbanization of American Physics." *American Quarterly* 56, no. 4 (December 2004), 851–88.

Kammen, Michael G. *Mystic Chords of Memory: The Transformation of Tradition in American Culture*. New York: Knopf, 1991.

Kawin, Ethel. *The Wise Choice of Toys*. Chicago: University of Chicago Press, 1938.

Keene, Melanie. "'Every Boy and Girl a Scientist': Instruments for Children in Interwar Britain." *Isis* 98, no. 2 (June 1, 2007): 266–89.

———. *Science in Wonderland: The Scientific Fairy Tales of Victorian Britain*. Oxford: Oxford University Press, 2015.

Keeney, Elizabeth. *The Botanizers: Amateur Scientists in Nineteenth-Century America*. Chapel Hill: University of North Carolina Press, 1992.

Ketcham, Hank. *Dennis the Menace*, no. 135. New York: Fawcett, 1974.

Kevles, Daniel J. *The Physicists: The History of a Scientific Community in Modern America*. New York: Knopf, 1978.

Kidd, Kenneth B. *Making American Boys: Boyology and the Feral Tale*. Minneapolis: University of Minnesota Press, 2004.

Klapper, Melissa R. *Small Strangers: The Experiences of Immigrant Children in America, 1880–1925*. Chicago: Ivan R. Dee, 2007.

Knapp, R. H., and Wesleyan University (Middletown, Conn.). *Origins of American Scientists: A Study Made under the Direction of a Committee of the Faculty of Wesleyan University*. Chicago: University of Chicago Press for Wesleyan University, 1952.

Kohler, Robert E. *All Creatures: Naturalists, Collectors, and Biodiversity, 1850–1950*. Princeton, N.J.: Princeton University Press, 2006.

Kohlstedt, Sally Gregory. "Parlors, Primers, and Public Schooling: Education for Science in Nineteenth-Century America." *Isis* 81, no. 3 (September 1990): 424–45.

———. *Teaching Children Science: Hands-On Nature Study in North America, 1890–1930*. Chicago: University of Chicago Press, 2010.

Kramer, Rita. *Maria Montessori: A Biography*. New York: Putnam, 1976.

Kuznick, Peter J. "Losing the World of Tomorrow: The Battle over the Presentation of Science at the 1939 New York World's Fair." *American Quarterly* 46, no. 3 (1994): 341–73.

Ladd-Taylor, Molly, and Lauri Umansky, eds. *"Bad" Mothers: The Politics of Blame in Twentieth-Century America*. New York: New York University Press, 1998.

LaFollette, Marcel C. *Making Science Our Own: Public Images of Science, 1910–1955.* Chicago: University of Chicago Press, 1990.

———. *Science on the Air: Popularizers and Personalities on Radio and Early Television.* Chicago: University of Chicago Press, 2008.

Leach, William. *Butterfly People: An American Encounter with the Beauty of the World.* New York: Pantheon Books, 2013.

Lecklider, Aaron. *Inventing the Egghead: The Battle over Brainpower in American Culture.* Philadelphia: University of Pennsylvania Press, 2013.

Lederer, Susan E. *Subjected to Science: Human Experimentation in America before the Second World War.* Baltimore: Johns Hopkins University Press, 1995.

Lemann, Nicholas. *The Big Test: The Secret History of the American Meritocracy.* New York: Farrar, Straus and Giroux, 1999.

Leslie, Stuart W. *The Cold War and American Science: The Military-Industrial-Academic Complex at MIT and Stanford.* New York: Columbia University Press, 1993.

Lewenstein, B. V. "The Meaning of 'Public Understanding of Science' in the United States after World War II." *Public Understanding of Science* 1, no. 1 (1992): 45–68.

Lienhard, John H. *Inventing Modern: Growing Up with X-Rays, Skyscrapers, and Tailfins.* New York: Oxford University Press, 2005.

Louv, Richard. *Last Child in the Woods: Saving Our Children from Nature-Deficit Disorder.* Chapel Hill, N.C.: Algonquin Books, 2008.

Luey, Beth. *Expanding the American Mind: Books and the Popularization of Knowledge.* Amherst: University of Massachusetts Press, 2010.

Macleod, David I. *Building Character in the American Boy: The Boy Scouts, YMCA, and Their Forerunners, 1870–1920.* Madison: University of Wisconsin Press, 1983.

Marchand, Roland. *Advertising the American Dream: Making Way for Modernity, 1920–1940.* Berkeley: University of California Press, 1985.

Marcus, Leonard S. *Minders of Make-Believe: Idealists, Entrepreneurs, and the Shaping of American Children's Literature.* Boston: Houghton Mifflin, 2008.

Marsden, George M. *Fundamentalism and American Culture.* New York: Oxford University Press, 2006.

Martin, Bill, Jr., and Jay Ells. *Spoiled Tomatoes.* Glendale, Calif.: Bowman, 1970.

Mason, Jennifer. *Civilized Creatures: Urban Animals, Sentimental Culture, and American Literature, 1850–1900.* Baltimore: Johns Hopkins University Press, 2005.

Matt, Susan J. "Children's Envy and the Emergence of the Modern Consumer Ethic, 1890–1930." *Journal of Social History* 36, no. 2 (Winter 2002): 283–302.

McCurdy, Howard E. *Space and the American Imagination.* Washington, D.C.: Smithsonian Institution Press, 1997.

Medovoi, Leerom. *Rebels: Youth and the Cold War Origins of Identity.* Durham, N.C.: Duke University Press, 2005.

Mendlesohn, Farah. *The Inter-galactic Playground: A Critical Study of Children's and Teens' Science Fiction.* Jefferson, N.C.: McFarland, 2009.

Mickenberg, Julia L. *Learning from the Left: Children's Literature, the Cold War, and Radical Politics in the United States*. New York: Oxford University Press, 2006.

Miller, Cynthia J., and A. Bowdoin Van Riper, eds. *1950's "Rocketman" TV Series and Their Fans: Cadets, Rangers, and Junior Space Men*. New York: Palgrave Macmillan, 2012.

Miller, Jon D. "Public Understanding of, and Attitudes toward, Scientific Research: What We Know and What We Need to Know." *Public Understanding of Science* 13, no. 3 (2004): 273–94.

Mitchell, Lucy Sprague. *Here and Now Story Book, Two-to-Seven-Year-Olds: Experimental Stories Written for the Children of the City and Country School (Formerly the Play School) and the Nursery School of the Bureau of Educational Experiments*. New York: E. P. Dutton, 1921.

Montoya, Fawn-Amber. "Model Schools and Field Days: Colorado Fuel and Iron's Construction of Education and Recreation for Children, 1901–1918." In *Children and Youth during the Gilded Age and Progressive Era*, edited by James Alan Marten, 42–58. New York: New York University Press, 2014.

Morus, Iwan Rhys. "Seeing and Believing Science." *Isis* 97, no. 1 (March 2006): 101–10.

———. "Worlds of Wonder: Sensation and the Victorian Scientific Performance." *Isis* 101, no. 4 (December 2010): 806–16.

Muchnick, Barry Ross Harrison. "Nature's Republic: Fresh Air Reform and the Moral Ecology of Citizenship in Turn of the Century America." Ph.D. diss., Yale University, 2010. Accessed June 26, 2015. http://gradworks.umi.com/34/40/3440577.html.

Nadis, Fred. *Wonder Shows: Performing Science, Magic, and Religion in America*. New Brunswick, N.J.: Rutgers University Press, 2005.

Nathanson, Iric. "The Oppenheimer Affair: Red Scare in Minnesota." *Minnesota History* 60, no. 5 (April 1, 2007): 172–86.

The New Wonder World: Library of Knowledge. Chicago: Geo. L. Shuman, 1943.

Norman, Ralph David. "A Study of Scientifically Talented Boys, with Special Reference to the Early Validity of Selections Made in the First Science Talent Search." *Abstracts of Doctoral Dissertations* 52 (n.d.): 233–38.

O'Connor, Ralph. *The Earth on Show: Fossils and the Poetics of Popular Science, 1802–1856*. Chicago: University of Chicago Press, 2007.

Ogata, Amy F. "Building Imagination in Postwar American Children's Rooms." *Studies in the Decorative Arts* 16, no. 1 (2008): 126–42.

———. "Creative Playthings: Educational Toys and Postwar American Culture." *Winterthur Portfolio* 39, no. 2/3 (June 1, 2004): 129–56.

———. *Designing the Creative Child: Playthings and Places in Midcentury America*. Minneapolis: University of Minnesota Press, 2013.

Oldenziel, Ruth. "Boys and Their Toys: The Fisher Body Craftsman's Guild, 1930–1968, and the Making of a Male Technical Domain." In *Boys and Their Toys: Masculinity, Class, and Technology in America*, edited by Horowitz, Roger, 131–68. New York: Routledge, 2001.

Onion, Rebecca. "Honey, You're Scaring the Kids." *Appendix*, July 22, 2014. Accessed November 16, 2015. http://theappendix.net/issues/2014/7/honey-youre-scaring-the-kids.

———. "Writing a Wonderland of Science: Child-Authored Periodicals at the Brooklyn Children's Museum, 1936–1946." *American Periodicals* 23, no. 1 (2013): 1–21.

Onion, Rebecca Stiles. "Picturing Nature and Childhood at the American Museum of Natural History and the Brooklyn Children's Museum, 1899–1930." *Journal of the History of Childhood & Youth* 4, no. 3 (Fall 2011): 434.

Op de Beeck, Nathalie. *Suspended Animation: Children's Picture Books and the Fairy Tale of Modernity*. Minneapolis: University of Minnesota Press, 2010.

Pandora, Katherine. "The Children's Republic of Science in the Antebellum Literature of Samuel Griswold Goodrich and Jacob Abbott." *Osiris* 24, no. 1 (January 1, 2009): 75–98.

Paris, Leslie. *Children's Nature: The Rise of the American Summer Camp*. New York: New York University Press, 2008.

Parnall, Peter. *The Mountain*. Garden City, N.Y.: Doubleday, 1971.

Patterson, William H. *Robert A. Heinlein: In Dialogue with His Century*. Vol. 1, *1907–1948: Learning Curve*. New York: Tor Books, 2010.

Peet, Bill. *The Wump World*. Boston: Houghton Mifflin, 1970.

Perkins, Maureen. *The Reform of Time: Magic and Modernity*. London: Pluto Press, 2001.

Petersham, Maud Fuller, and Miska Petersham. *The Story Book of Gold*. Philadelphia: J. C. Winston, 1935.

Phares, Tom K., and Westinghouse Electric Corporation. *Seeking—and Finding—Science Talent: A 50-Year History of the Westinghouse Science Talent Search*. [Pittsburgh?]: Westinghouse Electric Corporation, 1990.

Pindar, George N. *Guide to the Nature Treasures of New York City: American Museum of Natural History, New York Aquarium, New York Zoölogical Park and Botanical Garden, Brooklyn Museum, Botanic Garden and Children's Museum*. New York: Published for the American Museum of Natural History by C. Scribner's Sons, 1917.

Plevin, Arlene. "Still Putting Out 'Fires': Ranger Rick and Animal/Human Stewardship." In *Wild Things: Children's Culture and Ecocriticism*, edited by Sidney I. Dobrin and Kenneth B. Kidd, 168–82. Detroit: Wayne State University Press, 2004.

Pollack, Eileen. *The Only Woman in the Room: Why Science Is Still a Boys' Club*. Boston: Beacon Press, 2015.

Popkewitz, Thomas S. *Inventing the Modern Self and John Dewey: Modernities and the Traveling of Pragmatism in Education*. New York: Palgrave Macmillan, 2005.

Price, Jennifer. *Flight Maps: Adventures with Nature in Modern America*. 1st ed. New York: Basic Books, 1999.

Pryor, William Clayton, and Helen Sloman Pryor. *The Steel Book: A Photographic Picture-Book with a Story*. New York: Harcourt, Brace, 1935.

———. *The Streamline Train Book: A Photographic Picture-Book with a Story*. New York: Harcourt, Brace, 1937.

Puaca, Laura Micheletti. *Searching for Scientific Womanpower: Technocratic Feminism and the Politics of National Security, 1940–1980*. Chapel Hill: University of North Carolina Press, 2014.

Pursell, Carroll W. "Toys, Technology, and Sex Roles in America, 1920–1940." In *Dynamos and Virgins Revisited: Women and Technological Change in History*, edited by Martha Moore Trescott, 252–67. Metuchen, N.J.: Scarecrow, 1979.

Rader, Karen A., and Victoria E. M. Cain. "From Natural History to Science: Display and the Transformation of American Museums of Science and Nature." *Museum & Society* 6, no. 2 (July 2008): 152–71.

———. *Life on Display: Revolutionizing U.S. Museums of Science and Natural History in the Twentieth Century*. Chicago: University of Chicago Press, 2014.

Rae, John. "Application of Science to Industry." In *The Organization of Knowledge in Modern America, 1860–1920*, edited by Alexandra Oleson, 249–69. Baltimore: Johns Hopkins University Press, 1979.

Rahn, Suzanne. "'Like a Star through Flying Snow': Jewish Characters, Visible and Invisible." *The Lion and the Unicorn* 27, no. 3 (2003): 303–23.

Rauch, Alan. "A World of Faith on a Foundation of Science: Science and Religion in British Children's Literature, 1761–1878." *Children's Literature Association Quarterly* 14, no. 1 (1989): 13–19.

Redcay, Anna M. "'Live to Learn and Learn to Live': The St. Nicholas League and the Vocation of Childhood." *Children's Literature* 39 (2011): 58–84.

Reich, Leonard S. "From the Spirit of St. Louis to the SST: Charles Lindbergh, Technology, and Environment." *Technology and Culture* 36, no. 2 (April 1, 1995): 351–93.

Rhees, David. "The Chemists' Crusade: The Rise of an Industrial Science in Modern America, 1907–1922." Ph.D. diss., University of Pennsylvania, 1987.

———. "A New Voice for Science: Science Service under Edwin E. Slosson, 1921–1929." Master's thesis, University of North Carolina at Chapel Hill, 1979.

Rickover, Hyman. *Education and Freedom*. New York: Dutton, 1959.

Roe, Anne. *The Making of a Scientist*. New York: Dodd, Mead, 1953.

———. "Personal Problems and Science." In *Scientific Creativity, Its Recognition and Development. Selected Papers from the Proceedings of the First, Second, and Third University of Utah Conferences: The Identification of Creative Scientific Talent*, edited by Calvin W. Taylor and Frank Barron, 132–38. New York: Wiley, 1963.

Rome, Adam. *The Genius of Earth Day: How a 1970 Teach-in Unexpectedly Made the First Green Generation*. New York: Hill and Wang, 2013.

Rose, Jacqueline. *The Case of Peter Pan, or, The Impossibility of Children's Fiction*. London: Macmillan, 1984.

Rossiter, Margaret W. *Women Scientists in America: Before Affirmative Action, 1940–1972*. Baltimore: Johns Hopkins University Press, 1995.

———. *Women Scientists in America: Struggles and Strategies to 1940*. Baltimore: Johns Hopkins University Press, 1982.

Rouyer, Anne. "How Did YA Become YA?" *Stuff for the Teen Age*. Accessed June 4, 2015. http://www.nypl.org/blog/2015/04/20/how-did-ya-become-ya.

Rubin, Joan Shelley. *The Making of Middlebrow Culture*. Chapel Hill: University of North Carolina Press, 1992.

Rudolph, John L. *Scientists in the Classroom: The Cold War Reconstruction of American Science Education*. New York: Palgrave, 2002.

———. "Turning Science to Account: Chicago and the General Science Movement in Secondary Education, 1905–1920." *Isis* 96, no. 3 (2005): 353–89.

Russ, Joanna. "Images of Women in Science Fiction." In *Images of Women in Fiction: Feminist Perspectives*, edited by Susan Koppelman Cornillon, 79–94. Bowling Green, Ohio: Bowling Green University Popular Press, 1973.

Ryan, Cornelius, ed. *Across the Space Frontier*. New York: Viking, 1952.

Rydell, Robert W. *World of Fairs: The Century-of-Progress Expositions*. Chicago: University of Chicago Press, 1993.

Sabin, Paul. *The Bet: Paul Ehrlich, Julian Simon, and Our Gamble over Earth's Future*. New Haven: Yale University Press, 2013.

Sacks, Oliver W. *Uncle Tungsten: Memories of a Chemical Boyhood*. New York: Alfred A. Knopf, 2001.

Sammond, Nicholas. *Babes in Tomorrowland: Walt Disney and the Making of the American Child, 1930–1960*. Durham, N.C.: Duke University Press, 2005.

Sánchez-Eppler, Karen. "Castaways: The Swiss Family Robinson, Child Bookmakers, and the Possibilities of Library Flotsam." In *The Oxford Handbook of Children's Literature*, edited by Julia L. Mickenberg and Lynne Vallone, 433–54. New York: Oxford University Press, 2011.

———. *Dependent States: The Child's Part in Nineteenth-Century American Culture*. Chicago: University of Chicago Press, 2005.

Schlereth, Thomas J. *Cultural History and Material Culture: Everyday Life, Landscapes, Museums*. Charlottesville: University Press of Virginia, 1992.

Schofield-Bodt, Cindy. "A History of Children's Museums in the United States." *Children's Environments Quarterly* 4, no. 1 (Spring 1987): 4–6.

Scripps, E. W., and Charles R. McCabe. *Damned Old Crank: A Self-Portrait of E. W. Scripps Drawn from His Unpublished Writings*. New York: Harper and Brothers, 1951.

Secord, James A. "Knowledge in Transit." *Isis* 95, no. 4 (December 2004): 654–72.

———. *Victorian Sensation: The Extraordinary Publication, Reception, and Secret Authorship of "Vestiges of the Natural History of Creation."* Chicago: University of Chicago Press, 2000.

Shapin, Steven. *The Scientific Life: A Moral History of a Late Modern Vocation*. Chicago: University of Chicago Press, 2008.

Slosson, Edwin E. *Creative Chemistry, Descriptive of Recent Achievements in the Chemical Industries*. New York: Appleton-Century, 1938.

Slotten, Hugh R. "Humane Chemistry or Scientific Barbarism? American Responses to World War I Poison Gas, 1915–1930." *Journal of American History* 77, no. 2 (September 1, 1990): 476–98.

Smith, John Kenly. "The Evolution of the Chemical Industry: A Technological Perspective." In *Chemical Sciences in the Modern World*, edited by Seymour H. Mauskopf, 137–57. Philadelphia: University of Pennsylvania Press, 1993.

Solovey, Mark. "Cold War Social Science: Specter, Reality, or Useful Concept?" In *Cold War Social Science: Knowledge Production, Liberal Democracy, and Human Nature*, edited by Mark Solovey and Hamilton Cravens, 1–24. Basingstoke, England: Palgrave Macmillan, 2012.

Spigel, Lynn. *Welcome to the Dreamhouse: Popular Media and Postwar Suburbs*. Durham, N.C.: Duke University Press, 2001.

Spring, Joel H. *The Sorting Machine: National Educational Policy since 1945*. New York: McKay, 1976.

Stearns, Peter N. *Anxious Parents: A History of Modern Childrearing in America*. New York: New York University Press, 2003.

———. "Obedience and Emotion: A Challenge in the Emotional History of Childhood." *Journal of Social History* 47, no. 3 (March 1, 2014): 593–611.

Stephens, Sharon. *Children and the Politics of Culture*. Princeton, N.J.: Princeton University Press, 1995.

Sullivan, C. W., III. "American Young Adult Science Fiction since 1947." In *Young Adult Science Fiction*, edited by C. W. Sullivan III, 21–35. Westport, Conn.: Greenwood Press, 1999.

———. "Heinlein's Juveniles: Growing Up in Outer Space." In *Science Fiction for Young Readers*, edited by C. W. Sullivan III, 21–35. Westport, Conn.: Greenwood Press, 1993.

———. "Heinlein's Juveniles: Still Contemporary after All These Years." *Children's Literature Association Quarterly* 10, no. 2 (2009): 64–66.

Svilpis, J. "Authority, Autonomy, and Adventure in Juvenile Science Fiction." *Children's Literature Association Quarterly* 8, no. 3 (2009): 22–26.

Syder, Andrew Derek. "'Shaken Out of the Ruts of Ordinary Perception': Vision, Culture and Technology in the Psychedelic Sixties." Ph.D. diss., University of Southern California, 2009.

Tarr, Joel, and Mark Tebeau. "Housewives as Home Safety Managers: The Changing Perception of the Home as a Place of Hazard and Risk, 1870–1940." In *Accidents in History: Injuries, Fatalities and Social Relations*, edited by Roger Cooter and Bill Luckin, 196. Amsterdam: Editions Rodopi B.V., 1997.

———. "Managing Danger in the Home Environment, 1900–1940." *Journal of Social History* 29, no. 4 (July 1, 1996): 797–816.

Tebbel, John. "For Children, with Love and Profit: Two Decades of Book Publishing for Children." In *Stepping Away from Tradition: Children's Books of the Twenties and Thirties*, edited by Sybille A. Jagusch. Washington, D.C.: Library of Congress, n.d.

Teitelbaum, Michael S. *Falling Behind? Boom, Bust, and the Global Race for Scientific Talent*. Princeton, N.J.: Princeton University Press, 2014.

Terry, Jennifer. "'Momism' and the Making of Treasonous Homosexuals." In *"Bad" Mothers: The Politics of Blame in Twentieth-Century America*, edited by Molly Ladd-Taylor and Lauri Umansky, 169–90. New York: New York University Press, 1998.

Terzian, Sevan. "'Adventures in Science': Casting Scientifically Talented Youth as National Resources on American Radio, 1942–1958." *Paedagogica Historica* 44, no. 3 (June 2008): 309–25.

———. "The 1939–1940 New York World's Fair and the Transformation of the American Science Extracurriculum." *Science Education* 93, no. 5 (September 2009): 892–914.

———. *Science Education and Citizenship: Fairs, Clubs and Talent Searches for American Youth, 1918–1958*. New York: Palgrave Macmillan, 2013.

———. "Science World, High School Girls, and the Prospect of Scientific Careers, 1957–1963." *History of Education Quarterly* 46, no. 1 (2006): 73–99.

Tobey, Ronald C. *The American Ideology of National Science, 1919–1930*. Pittsburgh: University of Pittsburgh Press, 1971.

Tolley, Kimberley. *The Science Education of American Girls: A Historical Perspective*. New York: RoutledgeFalmer, 2003.

Tribbe, Matthew D. *No Requiem for the Space Age: The Apollo Moon Landings and American Culture*. New York: Oxford University Press, 2014.

Tyler, John. *The Chemcraft Story: The Legacy of Harold Porter*. Haworth, N.J.: St. Johann Press, 2003.

Unti, Bernard, and Bill DeRosa. "Humane Education Past, Present, and Future." In *The State of the Animals II: 2003*, edited by D. J. Salem and A. N. Rowan, 27–50. Washington, D.C.: Humane Society Press, 2003.

Valentine, Deborah. "Playing Progressively? Race, Reform, and Playful Pedagogies in the Origins of Philadelphia's Starr Garden Recreation Park, 1857–1904." In *Children and Youth during the Gilded Age and Progressive Era*, edited by James Alan Marten, 19–41. New York: New York University Press, 2014.

Wade, Nicholas J. "Philosophical Instruments and Toys: Optical Devices Extending the Art of Seeing." *Journal of the History of the Neurosciences* 13, no. 1 (March 2004): 102–24.

Walkerdine, Valerie. *The Mastery of Reason: Cognitive Development and the Production of Rationality*. London: Routledge, 1988.

Wang, Jessica. *American Science in an Age of Anxiety: Scientists, Anticommunism, and the Cold War*. Chapel Hill: University of North Carolina Press, 1999.

Wang, Zuoyue. *In Sputnik's Shadow: The President's Science Advisory Committee and Cold War America*. New Brunswick, N.J.: Rutgers University Press, 2008.

Wannamaker, Annette. *Boys in Children's Literature and Popular Culture: Masculinity, Abjection, and the Fictional Child*. New York: Routledge, 2007.

Waterman, Alan T. Introduction to *Science, the Endless Frontier: A Report to the President* on a Program for Postwar Scientific Research, by Vannevar Bush. Washington, D.C.: National Science Foundation, 1960.

Watson, Bruce. *The Man Who Changed How Boys and Toys Were Made*. New York, N.Y.: Viking, 2002.

West, Nancy Martha. *Kodak and the Lens of Nostalgia*. Charlottesville: University Press of Virginia, 2000.

Whyte, William Hollingsworth. *The Organization Man*. New York: Simon and Schuster, 1972.

Wise, Winifred Esther. *Young Edison: The True Story of Edison's Boyhood*. Chicago: Rand McNally, 1933.

Wylie, Philip. *Generation of Vipers*. New York: Farrar Rinehart, 1942.

Yarrow, Andrew L. *Thrift: The History of an American Cultural Movement*. Amherst: University of Massachusetts Press, 2014.

Youth Looks at Science and War. Washington, D.C.: Science Service and Penguin, 1942.

Zelizer, Viviana A. Rotman. *Pricing the Priceless Child: The Changing Social Value of Children*. New York: Basic Books, 1985.

Index